THE
ESTROGEN ALTERNATIVE

THE
ESTROGEN ALTERNATIVE

NATURAL HORMONE THERAPY WITH BOTANICAL PROGESTERONE

RAQUEL MARTIN
WITH JUDI GERSTUNG, D.C.

Healing Arts Press
Rochester, Vermont

Healing Arts Press
One Park Street
Rochester, Vermont 05767
www.gotoit.com

*Note to the reader: This book is intended as an informational guide. The remedies, approaches, and
techniques described herein are meant to supplement, and not to be a substitute for, professional
medical care or treatment. They should not be used to treat a serious ailment without prior
consultation with a qualified health-care professional.*

Library of Congress Cataloging-in-Publication Data
Martin, Raquel, 1935–
The estrogen alternative : natural hormone therapy with botanical progesterone /
Raquel Martin, with Judi Gerstung. — Rev. & expanded ed.
p.   cm.
Includes bibliographical references and index.
ISBN 0-89281-780-1 (pbk. : alk. paper)
1. Progesterone—Therapeutic use. 2. Generative organs, Female—Diseases—Alternative
treatment. 3. Menopause—Complications—Alternative treatment. 4. Plant hormones—
Therapeutic use. 5. Generative organs, Female—Diseases—Hormone therapy.
I. Gerstung, Judi. II. Title.
RG129.P66M37 1998
618.1'061—dc21   97-42690
                                                  CIP

Printed and bound in Canada

10 9 8 7 6 5 4 3 2

Text design and layout by Kristin Camp
This book was typeset in Caslon

Healing Arts Press is a division of Inner Traditions International

Distributed to the book trade in Canada by Publishers Group West (PGW),
Toronto, Ontario
Distributed to the health food trade in Canada by Alive Books, Toronto and Vancouver
Distributed to the book trade in the United Kingdom by Deep Books, London
Distributed to the book trade in Australia by Millennium Books, Newtown, N.S.W.
Distributed to the book trade in New Zealand by Tandem Press, Auckland
Distributed to the book trade in South Africa by Alternative Books, Ferndale

CONTENTS

FOREWORD TO THE SECOND EDITION

The 1990s have brought about considerable changes in health care in this country. Many of these changes result from the emergence of managed care organizations in an attempt to reduce medical costs. This approach, unfortunately, has brought with it limitations and restrictions on health care delivery. Many consumers are now turning to more user-friendly complementary and alternative care in which individuals take an active part in the decisions necessary to maintain optimum health.

Expanding knowledge of the usefulness of such complementary care is supported by the Dietary Supplement Health and Education Act of 1994. A major aspect of this includes the greater use of plant chemicals (phytochemicals) as dietary supplements or "medicinals," which in many cases may replace synthetic pharmaceuticals.

As women become more and more involved in decisions about hormone replacement therapy (HRT), *The Estrogen Alternative* serves a vital need. It is very timely and addresses this increasingly complex problem. The dilemma of HRT today stems largely from the many inconclusive and contradictory studies published by traditional medical institutions.

The woman who chooses to be an active participant in decisions regarding hormone therapy must first heed Raquel Martin's reminder that we still have a lot to learn. With that in mind, the author has presented an extensive amount of information on HRT, with emphasis on the use of natural progesterone. The decisions a woman makes must include consideration of risk versus benefit, a vital factor in all health-care choices. It's not an easy task, and unfortunately, the focus of most scientific studies has been only synthetic preparations. This book addresses the many aspects of HRT, whether natural or synthetic, that we need to consider. On one hand, benefits include a reduction in both osteoporosis (progesterone decreases the risk of bone fractures and vertebral body collapse) and vascular disease (it also decreases the risk of heart attacks and stroke). On the other hand is increased risk of breast and endometrial cancer with traditional HRT. A woman's quality of life, as well as her longevity, are influenced by her decisions about hormone therapy. Yet much of the information that reaches the general public provides inconclusive data on which to base proper decisions.

The Estrogen Alternative examines the benefits of natural progesterone therapy for women of all ages. The author's shared experiences make it even more appealing. She provides educational support for women who wish to participate in decisions about their own care. She also presents a challenge to the physicians who, it is hoped, will become more receptive to patients considering a more natural approach.

I. Sylvia Crawley, M.D.
Chair, Medical Education Committee
American Nutraceutical Association

FOREWORD TO THE FIRST EDITION

Many good books have been written about natural health alternatives, a large number of them by health-care professionals. Raquel Martin's well-written book is unique, however, in that it comes to us from a layperson—a consumer of health care, if you will. As a woman, she herself has had to cope with many of the problems described here. As a knowledgeable health activist and author of *Today's Health Alternative*, she has investigated the options open to her and examined those that were most effective. As a dedicated writer, she has researched her subject, gathered the important facts from scientific journals and the medical literature about the benefits of natural foods and hormones, and, finally, collaborated with health-care professionals to put it all together into an accurate and highly readable text.

This book will be important to women around the world who have been bombarded with chemical therapies. They have been on a physiological roller coaster, and it's time for them to have safe and effective solutions. Raquel Martin is an inspiring role model for all the women who are not getting the help they need. She urges them to seek out the answers for themselves, sort out for themselves the good from the bad, and find out what works and what doesn't—in short, to take charge of their own health care, be more informed about prescription and over-the-counter pharmaceuticals, and ask the right questions of their doctors.

My grandfather founded a pharmaceutical company more than a century ago. In the old days he and my father manufactured herbal products and other concentrates of botanical origin for physicians all over the country. As medicine began to change and drug companies decided to patent and market synthetic products instead, the natural remedies became scarce. I chose to carry on the family tradition by creating natural products that make a difference.

More than twenty years of studying phytogenins led me to effective natural sources for a number of hormones. Along the way I became impressed with the efficacy of the wild yam. I have devoted a lot of time to studying the yam and identifying the characteristics and bioavailability of different species. Wild yams are an unbelievably effective resource for natural hormone therapy if you select the right species and extraction methods. Let's hope that in the future these products will be standardized and properly labeled, so the consumer can know what she is getting.

When I learned about the effort that Raquel Martin has expended to find out everything she could from medical reports, medical doctors, natural health-care practitioners, pharmacies, manufacturers, and other studies on the effectiveness of the wild yam, and in particular phytogenin markers from specific species, I was intrigued. The anecdotal evidence from women who have been helped affirms the value of this botanical supplement.

I feel it is an honor to be part of this work, which brings a wealth of good information to the table for our consideration. Many doctors have written books that are biased toward drugs, and naturalists have written books slanted toward herbs. But this team of authors has addressed a subject of vital importance to women in a simple, straightforward manner—without bias, and with a lot of very useful information.

It's refreshing to read a book written from the soul. Perhaps someday the author's song will become a full chorus. Forget about the science for a moment; zero in on the message: Women have a choice! Take control of your health! Read the book, study the alternatives, ask questions, and above all, think for yourself.

James Jamieson
Pharmacologist

ACKNOWLEDGMENTS

This book was achieved through unmatched creative teamwork, where everyone's goal was the same: to raise public consciousness about disease prevention by exploring our choices and documenting the remarkable health benefits of natural alternatives. To Judi Gerstung, D.C., I owe untold thanks for the long hours she devoted to this undertaking—time away from her other jobs, as chiropractor, radiologist, teacher, and lecturer. Her comprehensive background in the health-care field has been invaluable. She assisted me in the organization of rapidly changing information and with editing throughout the project, making clarifying revisions in a logical and orderly way. Her desire to bring this immensely important subject to the attention of women worldwide and her special interest in the detection and prevention of osteoporosis guided me through tough and discouraging moments as we worked to reach our goal.

And my thanks to Irene Inglis, a long-time friend, for her support and partnership in our mutual quest for safe alternatives to drugs. The ideal combination of her training as a registered nurse with her interest in natural means of healing helped me transform medical terminology into practical knowledge. She was truly God-sent, with her assistance in proofreading and meeting necessary deadlines.

Appreciation goes to Laurie Skiba, writer and writing consultant, who contributed essential insight to some of the more difficult and technical chapters in the book's early stages. She generously shared her professional talent and experience, sacrificing time from her freelance work and the writing of her own books. Her critical analysis spurred me to deepen my investigation. I gained much from her intelligent, thought-provoking editorial comments.

Thank you to Marcia Jones, Director of the PMS and Menopause Center at Dixie Health, Inc., for her advice on the manuscript, for contributing new reference material, and for her many introductions to health-care advocates. Through her enthusiasm to get the facts out to women, I was introduced to Coeli Carr. I am sincerely grateful to both Coeli and Marcia for their generous counsel, as they guided me in my attempt to educate and motivate women toward a better quality of life.

Special thanks to Katharina Dalton, M.D.; Ray Peat, Ph.D.; Wallace Simons, R.Ph. (Women's International Pharmacy); Scott Stamper, R.Ph. (Women's Health America Group); Mark H. Mandel, R.Ph. (Snyder-Mark Drugs Roselle); Ole Krarp; Sam Georgiou, R.Ph. (Professional Arts Pharmacy); Brad Sorenson; and James Jamieson, pharmacologist, for supplying pertinent research data and documentation and helping me better appreciate the importance of natural HRT; and to Stuart and Jane Burke for testimonials that added a personal touch to this story. And thanks to Ray Silverman, Sasha Silverman, Debbie Dombrowski, Charis Dike, David Bundrick, Karla Emling, Beverly Kapple, Audrey Franklin, and Judy Sanger for their input and encouragement. I'll be forever grateful to J. K. Humber, Jr., D.C.; Mark Baker, D.C.; Judi Gerstung, D.C.; and Linda Force, D.C.; for providing case histories and moral support while also keeping me in good health during the stressful times.

And my love and thanks to my husband, Jack, for sharing his refreshing and clear communication skills. He furnished the missing "flow" to the work. His gift for words, analytical ability, and prudent judgment provided a consistent sounding board for my endless questions. He also helped me clarify my personal objectives and stay focused on them. He taught me necessary computer skills and usually stayed calm in the face of my recurring crises with so-called user-friendly software. His counsel held everything together.

And I can't thank Bill Martin, my brother-in-law, fervently enough for his suggestions and constructive criticism upon carefully reviewing an early draft. His objective eye zeroed in on better readability and accuracy, while his professional advice in the area of the sciences stimulated all of us to do more research.

Later came the difficult task of final revision and organization of the manuscript. For this and much more I am indebted to Connie Smith. Thanks to her painstaking regard for detail and priorities, our critical deadlines were all met. She brought to the book the accuracy and meticulous editing skills needed to do justice to a vital subject. Connie's personal research played an important role in unraveling and interpreting the differing viewpoints of various authorities. Her astute sense of judgment concerning complex issues was continually employed, as was her intelligent guidance. Her inspiration and strength motivated me to engage in honest reflection right to the end of our mission.

Last, I must acknowledge Dr. John R. Lee, whose research, personal practice, and writing have helped women to find more satisfying answers to age-old problems at long last. I am myself deeply grateful for his personal tutelage in scientific and medical terminology in preparation for this book's publication. I am honored that he agreed to become an active participant in this work. My thanks to Dr. Lee are echoed throughout the book.

And with both heart and mind I give thanks to the Lord for guiding me onto the paths of the above talented souls. I am grateful for His endowing us all with curiosity and concern about an issue so crucial, not only to our physical health, but also to our mental and spiritual well-being.

Dedicated to the doctors who are not limited by prejudice or greed, and who reason and act beyond the prevailing orthodox medical practice in order to pursue greater truth for the benefit of humanity; and to the multitudes of patients who suffer needlessly from having met only silence from the medical and pharmaceutical industries concerning the natural healing opportunities that have helped so many.

INTRODUCTION

Lead us onto the path of understanding new truths
and rendering a service to others.[1]

Helen Keller

In our lives we often face situations that are new, perhaps even overwhelming. We find we have to make educated choices without sufficient education being available to us. As women, we may be barred from the knowledge critical to managing our own unique health problems from puberty through old age. Natural hormone replacement therapy, for example, is seldom provided as an option, and few of us know about it. Instead, the medical world continues to stick with its same conventional therapies, exposing us to the hostile and long-term side effects of synthetic hormones.

This story begins years ago before much was known about the physiology of the menstural cycle. For many women the monthly cycle came and went with little or no discomfort. But for a significant number it was, as it is now, a time of physical and emotional affliction, often quite severe, and there was little that could be done about it.

Gradually, however, help began to arrive. The role of hormones in the menstrual cycle came to be understood better. Doctors began prescribing synthetic hormones as an antidote to the distress of premenstrual syndrome (PMS) and later in life's progression, menopause. Women were given the hope that these hormones would rid them of the unpleasant menstrual or menopausal symptoms that many of them experienced. But unfortunately, far too often, such hormones just made the symptoms worse.

Today, we have a fresh opportunity to halt the onslaught of hot flashes, night sweats, bloating, tension, fatigue, cramping, mercurial moods, and depression that may accompany an imbalance in our essential hormones. That opportunity is NHRT: natural hormone replacement therapy, as opposed to the usual synthetic HRT. I invite you to learn how and why plant-derived progesterone can help prevent the symptoms of PMS and menopause, including osteoporosis, fibrocystic breast disease, cardiovascular disorders, and painful endometriosis; how it can help with many pregnancy-related problems; and how it may help reverse diseases ranging from vaginal atrophy to heart disease and even cancer.[2, 3]

Part of our challenge is to find qualified, open-minded, good medical doctors to work with us. One of those high on my list is the late Dr. Robert S. Mendelsohn. His insight and spirit captured the hearts of many, and his philosophy echoes over and over again in a wealth of articles and books about health.

Besides drawing from the wisdom of Dr. Mendelsohn, I'm delighted to introduce to the reader several other eminent minds. One you will be hearing about in the course of this book is Julian Whitaker, M.D. Often, while browsing in health food stores, I've heard people talking about the latest breakthroughs in natural healing publicized in his newsletter. Many of these customers were at one time exhausted and disheartened by adverse reactions to the various drugs that had been prescribed to them. Natural options such as those presented by Dr. Whitaker help to ease the decisions we must sometimes make concerning our health.

After reading the following paragraph from his newsletter *Health & Healing*, I knew he would be added to my cadre of out-of-town medical advisors. He says, "In medical school . . . I was taught that the only tools that work to help people are drugs and surgery. In the twenty years since then, I have learned that a lot of what I was taught is just plain wrong. There are treatments for our most serious diseases that are not only safer than surgery and most drugs, but also *more powerful*." He continues, "The medical profession tends to promote what is good for the profession, not what is necessarily good for the patient. The doctor's joke, 'Better hurry up and operate before the patient gets well!' is truer than you think."[4]

I shudder to think of the patient who wants to make haste and "just get it over with" before looking into existing alternatives. I was such a patient many years ago. Now, shaped by life's experiences, I have learned to be much more inquisitive.

One day, reading one of Dr. Whitaker's newsletters, I enthusiastically jumped out of my chair and ran to tell my husband, Jack, about the abundance of information it gave on natural hormone replacement therapy. After having looked for so many years, I couldn't believe that such knowledge was suddenly available. I was so delighted I could hardly get the words out of my mouth. From this moment I was on the path that could possibly and finally resolve my longstanding misery—the path that would eventually lead to the publication of this book.

I soon became immersed in the findings of natural hormone research. I came across some incredible accounts, both personal and clinical, involving botanical progesterone.

The chronology of events begins in the year 1938 with the first adaptation of plant hormones to conform to human progesterone. In time, this discovery by research chemist Dr. Russell E. Marker[5] was followed by the work of other pioneers. Three decades later came Dr. Katharina Dalton's studies on the use of progesterone in prenatal care, recorded in the *British Journal of Psychiatry* (1968).[6] Ten years after that (1978), the *Journal of the American Medical Association* made further extraordinary revelations regarding natural hormone replacement;[7] and in 1989 the research of Joel T. Hargrove, M.D., and his colleagues on micronizing (finely grinding) progesterone for better absorption appeared in the *Journal of Obstetrics and Gynecology*.[8] A couple of years later (1991), a landmark review by Jerilynn C. Prior, M.D., on progesterone and the prevention of osteoporosis was published in the *Canadian Journal of OB/Gyn & Women's Health Care*.[9]

And for some of the most thorough and comprehensive work of the present day, we are indebted to John R. Lee, M.D., a leader in the field of natural progesterone. Raymond Peat, Ph.D. (an early mentor of Dr. Lee), Alan R. Gaby, M.D., Jonathan V. Wright, M.D., Betty Kamen, Ph.D., Christiane Northrup, M.D., and Jane Heimlich are some of the many others who have pursued the truth and have been willing to share it with all women.

The chapters to follow will emphasize the accumulated knowledge of these noted authorities on subjects that range from combating stress and lowering cholesterol to fertility, easier pregnancy, and successful breastfeeding. The last chapter of the book, as well as appendix F, is devoted to helping women locate a doctor interested in correcting the complex deficiencies that occur before, during, and after menopause.

My book was inspired by physicians who realize the importance of natural foods, vitamins, minerals, herbs, homeopathy, chiropractic, and, most specifically, botanical hormone replacement therapy. I hold in great esteem those who put themselves on the firing line, up against many of their colleagues, for the sake of what is best for their patients' health.

The health-care practitioners who will be featured in this book recognize the injustice the medical community has promoted in the name of "consumer welfare." Their compassionate efforts to reeducate the public have encouraged me to seek out, try, and then share these ideas. Although I wish I'd come across this information when I was younger, I feel blessed that it came my way at all. Now, past menopause, I actually feel more energetic than ever before. I have regained my health and vitality since discovering the natural way.

The revelations in these chapters should be read by all women and by those who care about them. We must start early to prevent chronic disease. Fatigue, headaches, heart disease, osteoporosis, and cancer, for example, can have various causes. Among these are a decline in hormones, unwholesome diet, food additives, environmental toxins, and nerve interference. Let's say, for our purposes, that the cause is indeed a hormone deficiency. If we take prescription drugs to cover up what is really a decline or lack of natural progesterone, more often than not our condition becomes worse.

Yet, the subject of hormone replacement is complicated because no two women's bodily needs are the same. For instance, past trauma or stress may cause some women to experience premenopausal symptoms in their thirties. Even the woman who sails through menopause without feeling any changes whatsoever is still at risk for cancer, heart disease, and osteoporosis. But regardless of how much estrogen we have, progesterone is our real concern because of the fact that very little is made anywhere in the body once ovulation ceases. As progesterone declines below one's estrogen level a hormone imbalance, with its many complications, is established.

Needless to say, it's important to know as much as possible about natural versus synthetic hormones and the side effects of the latter. First of all, you should know that *natural* progesterone is seldom prescribed by medical doctors. Yet, it was first crystallized from plant sources in 1938 and is readily available today. Most doctors don't know about it and don't even consider it when making recommendations to their patients. Why not? Probably because it is not a prescription drug. Although it may not be carried by your local pharmacist, it is available in specialized pharmacies (which are listed in appendix G) and health stores. However, most physicians haven't educated themselves or their patients about the benefits of such treatment.

John R. Lee, M.D., sheds some light on the reason why it's so hard to get information on natural progesterone from our traditional health-care providers:

> Pharmaceutical sales success and profit are . . . also dependent on the patentability of the compounds to be sold. Since natural compounds

(i.e., the hormone molecule as made by the ovary) cannot be patented, it is in the interest of the pharmaceutical industry to create compounds which are not identical to the natural hormone and are nowhere found in nature.[10]

Some medical doctors have told their patients that Provera, a synthesized compound, is progesterone. It is not![11] In fact, Dr. Lee informs us that it mimics only some of the activity of progesterone and is not identical. He says that altered and synthesized hormones such as Provera "may also provoke biologic responses which are undesirable or toxic. This is seen, for example, in the extensive lists of warnings, precautions, and side effects which accompany the descriptions of the synthetic hormones as found in the PDR (*Physicians' Desk Reference*)."[12] See appendix B.

Synthetic material does wonders in the making of cars, clothes, and Tupperware—but how desirable can it be for our biological needs? Before I started using natural hormones on a regular basis, I asked the pharmacists at the Women's International Pharmacy some questions. The research material they sent convinced me that natural progesterone is effective and is a plant-based substance containing no animal by-products. They explained that the botanical hormone diosgenin is a sterol, or saponin—an oil manufactured by many plants, including the wild yam, and easy to extract.

In the body, we make (synthesize) progesterone from cholesterol. In the lab, the chemist makes (synthesizes) progesterone from diosgenin. This *diosgenin* from the wild yam—like the stigmasterol from soybeans—is actually a precursor of a number of hormones. Thus, with only slight modification it can be made to duplicate the progesterone molecule the body produces so that it can be fully utilized as needed.[13]

The ensuing chapters will present a wealth of studies concerning the many applications of this natural hormone. We'll see how women who have suffered for most of their lives with various female problems have not only derived relief from their disorders but often achieved reversal of life-threatening illnesses such as osteoporosis and cancer.

We have no time to lose. I challenge women everywhere to use the information gathered here to evaluate for themselves all the pros and cons of hormonal treatment. We need to know the varied and multiple benefits of the natural form of progesterone over the synthetic, and to be thoroughly informed about the serious risks of pharmaceutical hormone products. As you will see in appendix G, even in the realm of natural hormone replacement therapy, there are wide variations in bioavailability and effectiveness in the multitude of new botanical products that are flooding the marketplace.

Unfortunately, previous generations have not had much choice in these matters. I can't help thinking of my own mother. Although she was a nurse, she didn't have this knowledge. I watched her slowly weaken from osteoporosis and disorders of the heart and kidneys. She became frail and bent over. Osteoporosis made her hip pain so severe that she was hardly able to lie on one side or even sit.

Because of my own experience with hormone imbalance, I have come to consider health trouble as a teacher, progressively opening my eyes to the power of the medicine found in nature. There is no need to be dependent on synthetic hormones or other drugs, which often cause sickness and premature aging. Age should not be a matter of how many years we have been living but rather a matter of the integrity of the tissues of the body. One's "age" is also shaped by a positive mental outlook—by seeking what is good and acting upon what is sound, not only for the body but for the mind and spirit.

This leads me to reflect upon the philosophy of a truly great woman who suffered from severe limitations. Helen Keller, rather than dwelling on her misfortune, actively sought ways in which she could be of some service to her fellow human beings as a contributing member of society. Her own constraints and her strong religious faith led Helen to understand and express in her writings a conflict we often face:

> We can drift along with general opinion and tradition, or we can throw ourselves upon the guidance of the soul within and steer courageously toward truth. . . . We have a choice in every event and every limitation and . . . to choose is to create.[14]

This book doesn't pretend to have all the answers, but it does provide a thought-provoking look at the choices now available. During the relatively short span of a few years that it took to compile and write it, the procession of ever-changing information has marched on. To the best of my ability, I have presented an accurate picture of natural hormones based on what we know today. Of course, some parts of this picture might change tomorrow.

But one thing we do know for certain is that what we *have* been doing in the way of hormone replacement is not working. We know there is a better way, and we've opened a dialogue that is bound to lead us there—as long as we continue to ask the right questions with open minds. Naturally, those at high risk or with a history of serious health problems should do their homework and ask even more questions. Here we've tried to initiate this process.

Seldom in the past have our medical options been made very clear to us; yet, these decisions can have such a profound influence on our lives. Since my venture into natural healing I have personally felt an exhilaration, never experienced before, about the choices that are now available among the more gentle forms of health care. As we realize and experience their harmony with nature, we will also discover a gift of life.

PART I

Natural Hormone Replacement from God's Garden

CHAPTER 1

SICK AND TIRED
OF BEING
TIRED AND SICK

Study sickness while you are well.
Thomas Fuller [1]

Something was wrong. I was in my forties, and the symptoms of menopause had already appeared. Hot flashes, bloating, irregularity, trouble
sleeping, night sweats, emotional tension—I realized that I needed some
kind of help to cope! The hormonal changes that every woman experiences
at midlife were playing havoc with my body. So I went to see my gynecologist. He responded with the standard, widely accepted medical treatment:
a program of hormone replacement therapy (HRT) using synthetic hormones. Specifically, this meant a prescription for Premarin (a conjugated
estrogen) and Provera (medroxyprogesterone acetate, a progestin). These
are manufactured substitutes for two hormones that play a central role in
every woman's sexual and reproductive life—hormones whose supply and
balance are gradually altered as a woman passes through and beyond the
normal childbearing years.

I had thought to myself, "Here I am in menopause. Finally, I'll have
freedom from my monthly periods." But instead I was told that I needed to
take these hormone supplements and continue drug-induced monthly
cycles for the rest of my life. For twenty-five days each month I was to take
one tablet of Premarin, and from the sixteenth through the twenty-fifth day
of each month I was to take the progestin tablet. Then I was to stop taking
both drugs for five days.

Even though I couldn't help feeling that my body was being artificially
regulated, my first reaction to the drugs was generally positive. My menstrual

cycle stabilized, the symptoms diminished, and I began to feel confident about my doctor's advice. By the second day my body was adjusting well. I became calmer and slowly began to feel better in many ways. "Why didn't I do this sooner?" I thought. "Such a simple solution to all of these problems!" I wanted to believe that routine HRT was the answer.

Before long, though, my honeymoon with synthetic hormones came to an end. In the second month I started to worry. There were some unpleasant new side effects, including weight gain, bloating, painful breasts, and tension. I wondered whether this was not the answer after all—at least, not the whole answer. Perhaps a change was in order. Maybe the dosage of the Premarin or Provera needed to be decreased or one or both of them discontinued, or maybe the whole approach was wrong. Disappointed now with my doctor's treatment, I consulted an endocrinologist, then an internist, and later still another gynecologist. All offered different suggestions, but none of their drugs helped my symptoms without bringing on some other abnormal discomfort. This trial-and-error period continued for several years.

Speaking with other women going through similar experiences, I heard a common complaint. They had gone down this same road with doctors who had prescribed variations of the same treatment, and as we all tend to do, followed their doctors' advice and just "hung in there." Women would come home from an appointment in tears of frustration because the doctor had made no change in their treatment regardless of the unpleasant reactions. The typical advice they got, as did I, was "You need to be patient. Just keep taking the pills a little longer until the body adapts." Generally a physician will urge a woman to continue her treatment, either varying the dosage or keeping it the same, on the premise that her body will eventually adjust to synthetic hormones. I can tell you that such advice will drive some women to their psychiatrists or closest medical centers in an attempt to deal with the drugs' multiple side effects.

I wanted my doctors to be right. So I did what they prescribed, over and over again. At first I'd be encouraged, because the drugs were making a few changes that I thought were good. But as time went on and it was clear that my body was reacting poorly, I began to feel additional symptoms: sharp uterine pain and inflammation and infection of the cervix, which was often quite painful. My bloating became more severe, along with digestive disorders such as colitis.

I feared another D & C (dilation and curettage), which some physicians routinely administer to women on HRT, and I dreaded the painful cauterization of cervical tissue that I'd been told was necessary to deal with the cervical inflammation. Yet I'd learned that the scar tissue resulting from some of the treatments I'd already had was probably causing even more harm to surrounding tissue! I grew more anxious still when a second endometrial biopsy (removal of a piece of the uterine lining by means of a plastic catheter) had to be performed in order to check for any deterioration of my uterus. Frustrated, I did not know what to do. I had to make up my mind before my next doctor's appointment, when he'd indicated he would probably recommend a hysterectomy.

As I continued to ask questions, I began to understand that the doctors I had been seeing all along didn't necessarily have all the answers. They didn't seem to comprehend completely the complexities of menopausal problems, the PMS problems of younger women, or the side effects of the synthetic hormones they were prescribing. Not only were their answers contradictory to each other; I sensed a lack of conviction on their part that this was indeed the right way to go. I thought to myself, *Is it any wonder women become confused, afraid, and discouraged during what can already be a stressful time?*

I decided I would have to take things into my own hands—maybe spend more time at the health store instead of the drugstore, and at least learn enough to direct my own treatment. I had lots of unanswered questions about why these hormone supplements were not working. What were synthetic hormones all about? Where did they come from and how did they work? What were the side effects—both immediate and long-term? Were alternative treatments available? What specific nutrients and exercises would help? And of course, the question that hung over it all: Whose advice should I believe, and what should I do?

Thus began my journey of discovery into the world of estrogen and progesterone and the roles these hormones play in women's health—a journey that not only led me to the answers to my many questions but in the process prepared me to recognize the variety of available alternatives. The answer has restored peace and health to my life.

THE SEARCH BEGINS

My goal became to find a complete and sound alternative treatment for my menopausal problems. After consultation with my doctor I immediately stopped using the synthetic progesterone substitute because it was causing dreadful feelings of stress throughout my body. I accepted his advice on the estrogen supplement, though, and continued to take it.

But I also began looking into other measures I had heard about, such as nutrition, herbs, and homeopathic remedies recommended specifically for menopause. I changed my diet to include more raw vegetables, fresh fruits, whole grains, seeds, and complex carbohydrates and less meat. These changes helped reduce some of my discomfort. I also added to my diet some known antioxidants (beta carotene; vitamins A, C, and E; and selenium) as well as vitamin B complex and zinc. I learned that zinc plays an important role in enzyme activity, especially in relation to the lymphocytes, and is needed for the absorption of vitamin A.[2] I had also become aware that many studies have shown that antioxidants help protect the body from toxins.

As long as I kept up my chiropractic treatments to enhance nerve flow and transmission, I was much less susceptible to gynecological infections and other disorders. I had learned that even though the right diet is important, the nervous system is the major regulator for all hormonal performance. I believe that spinal misalignments and subsequent nerve interference should always be addressed because of the close interconnection between the spinal nerves and the endocrine system.[3]

But those were the early years. Once the full force of menopause came upon me, some of the old symptoms resurfaced—though they were less pronounced than before. So either something major was still missing, or what I was taking (or doing) needed to be changed. I was ready to try anything just to feel normal. I talked to many other women in the same predicament, and eventually one of them gave me a new idea. She was a former nurse and knew about the Estraderm patch, a "time-release" patch that is placed on the surface of the skin so that estrogen is absorbed in small doses over time. I wondered why my doctor hadn't mentioned this.

I asked my doctor to substitute this method for my oral Premarin, and he agreed. This small adhesive pad did, in fact, seem to work better for me initially. Certain symptoms, such as the joint pains I'd been experiencing, temporarily went away. I felt more energetic at first, but I knew we hadn't found the solution yet.

I had read that estrogen should never be taken alone, so I sought out a female gynecologist and asked her about combining it with progesterone to cancel out the carcinogenic effect of the estrogen. When I told her I'd had a very bad reaction to Provera, she prescribed another of the synthetic substitutes for progesterone. Only later did I learn of the incredible difference between synthetic progesterone and real, natural progesterone.

Before long I again lapsed into some of the same symptoms, but this time it was from the effects of the synthetic hormones: nervousness, bloating, uterine cramping, and some sleepless nights. So I went to another doctor, and he decided to reduce the dose of estrogen to the lowest level available in patch form. As the months went by it became apparent that this wasn't working either. I knew something still wasn't right and was disappointed with all the experimenting. The vaginal dryness that had troubled me, before I began HRT, returned with all its associated pain and discomfort. I had also developed strange pains on the side of my breast, near the lymph nodes.

The doctor switched me to yet another synthetic progesterone substitute in the lowest dosage available. There followed more adjustments: less estrogen, then more, then back to Provera, then less of it, and so on. But nothing seemed to help. I kept trying in vain to find the right balance between the two drugs, thinking that the doctor knew best. I did not understand why my body was reacting as it did to these substances, or why they seemed to be doing more harm than good as time progressed.

I decided one day to stop the therapy completely. This decision came after I got out my magnifying glass and read the fine print that presented the risks and warnings on the leaflets accompanying the drugs. Some possible adverse reactions to synthetic hormones are liver disease, malignancy of the breast or genital organs, fluid retention, cystitis-like syndrome, headaches, nervousness, dizziness, edema, mental depression, insomnia, fatigue, and backache. The warnings on the package go on to point out that the drugs can cause or aggravate conditions such as epilepsy, migraine, asthma, and cardiac or renal dysfunction.[4] And all of the other estrogens and progestins normally prescribed have similar lists of side effects.

No wonder I was feeling bad. I began to resent the fact that the knowledge I desperately needed on such important health matters is not made readily available to women. At times I wondered how I would ever get through postmenopause if I couldn't get a grip on these menopausal years. Frankly, I was now afraid to take estrogen and the progesterone substitutes, which I understood at the time to be the only available source of progesterone.

But admittedly, I was getting some benefit from these drugs. When I stopped taking them, the old symptoms flared up. I even noticed that my chiropractic adjustments held better when I was on HRT. I also experienced an interesting phenomenon during these years of off-again, on-again hormone replacement: every time I stopped taking my hormones, I felt joint aches and knee pains. The more I monitored my hormone therapy by such signs and symptoms, the more suspicious I became that hormones (or the lack thereof) might very well be contributing to this reaction. I have since learned that other women have also come to associate muscle and joint pains with menopause.

On one hand, not taking the drugs was an obvious way to avoid the adverse reactions they could cause; on the other hand, giving them up meant losing their temporary benefits. A constant battle raged within my mind and body. I went back on the Estraderm patch, but in the meantime started reading everything I could find on the subject of hormones to try to find a new direction.

My hopes were turned into real fear when I read this statement in a report by Dr. Brian Henderson: "The patch produces higher levels of the most potent form of estrogen (estradiol) than does Premarin, giving a woman almost as much hormone as she would have made herself." Dr. Henderson continued, "The effect of that should be to make one's breast cancer risk go up substantially more on the patch than on Premarin."[5] It took that warning of a worst-case scenario for me to end my nearly ten years of experimenting with synthetic hormone replacement. I never found any combination or dosage of synthetic hormones that gave me enough benefits to compensate for the side effects that always accompanied their use. I felt blessed that I had not yet contracted cancer, but I remained terribly confused. Where could I turn next?

The beginning of the solution came to me when Dr. Julian Whitaker's *Health & Healing* newsletter arrived. What he had to say sent my hopes soaring. The information and phone numbers he included provided me answers to some of the questions I had asked doctors for so many years: "What specifically does *progesterone* do? How can we get natural hormones? Do we always need a prescription?" Could it be that what I was about to explore would actually resolve what had seemed a never-ending quest? Was it possible that my fears and anxieties could be a thing of the past? I truly hoped so.

My mailbox soon began to fill up with research reports and abstracts that I had requested. They told me that there was a natural, plant-based source of progesterone that had none of the side effects of the synthetic substitutes. Eager and willing, I kept rereading the encouraging words of Dr. John R. Lee from a statement in the periodical *Medical Hypotheses*: "Progesterone is inexpensive, being available from many plant sources. . . . Furthermore, it is remarkably free from side-effects."[6]

More information arrived in an educational brochure distributed by the Women's Health Connection in Madison, Wisconsin:

> The process of producing natural progesterone, which is made from yams and soybeans, was discovered by Russell Marker, a Pennsylvania State College chemistry professor. While experimenting with

sapogenins, a group of plant steroids, in the jungles of Mexico in the 1930s, Marker realized that progesterone could be transformed by chemical process from the sapogenin, diosgenin, which is found naturally in yams.

Unlike medroxyprogesterone [the chemical name for Provera], natural micronized progesterone is an exact chemical duplicate of the progesterone that is produced by the human body.

Another immediate difference between medroxyprogesterone and natural progesterone is that the synthetic hormone can actually lower a patient's blood level of progesterone. Some women who take medroxyprogesterone to combat PMS or oppose estrogen in menopause, report headaches, mood swings and fluid retention.

The more information I acquired, the more questions I had. How, why, and where does natural progesterone work in the body? The answers later came to fill a whole chapter of this book. But I am getting ahead of myself. I was so relieved to find that such a natural hormone existed that I immediately ordered a jar of cream made from these naturally occurring plant sterols. No prescription was needed. When it arrived, I quickly read the directions and applied the cream to my skin. It is a fat-soluble compound that is absorbed into the skin and taken up by the fatty layer beneath. From there it is transferred into the bloodstream to circulate throughout the body.

In the days and months that followed, I experienced a calmness and peace that I had not felt in many years. Enjoying my new sense of well-being was like living in a new body—no sharp uterine pains, no bloating, no tension. Best of all, my energy level was high, and I was able to sleep at night. I continued using the progesterone every day, secure in the knowledge that it was safe. It brought innumerable health benefits without artificially continuing my menstrual periods for the rest of my life.

I quickly learned that with *natural* hormone replacement therapy (NHRT), it was important for me to use the product daily for three weeks out of every month. I added this to my regimen of vitamins, minerals, other nutritional supplements, chiropractic care, and exercise. I found that all of these components were vital for optimal health and in supporting my natural HRT program. The plant-derived progesterone had greatly reduced the irritating conditions I've already described, such as fluid retention, colitis, joint pain, and sleep disorders. As my discomforts slowly diminished, I realized that this remarkable "phytohormone" had eliminated much of what used to be stress and had given me new stamina and energy.

As I gradually became more active I also found a peace of mind that is hard to describe. The benefits for me were obvious very quickly. Like most women I have talked to, I found that applying the contents of approximately one two-ounce jar of progesterone cream each month to various areas of the skin is quite sufficient to remedy the majority of problems. Further details on when and how to use the cream will be offered throughout the book; for other modes of hormone use, see appendix A.

Indeed, I have learned that progesterone can be found not only in the cream form but also in a capsule form (micronized for better absorption). I had my internist call in a prescription for a specific dosage of this to one of the numerous pharmacies that specialize in formulating natural products. At prescription strength, it is covered by most insurance companies, which makes NHRT quite affordable. The suppliers are located in various cities throughout the country (see appendix G). You can also order sublingual drops (applied under the tongue) or a micronized spray (applied to the mucous membranes of the cheeks) rather than the cream form. Some studies have found that the sublingual method provides approximately three times the concentration found in some of the nonprescription creams.[7]

I added this prescription for micronized capsules to my natural hormone program for two simple reasons: (1) I knew exactly how many milligrams I was taking, and (2) it was covered by insurance. At the time, I didn't want to bother with the blood or saliva testing because of the inconvenience and the expense. My choice served me well for several years, and I had never felt better.

As I read and learned more, however, I realized that although the prescription indicated exactly how many milligrams I was ingesting, this was not necessarily the amount of progesterone that was being absorbed by my body. Questions arose for me, such as, "After progesterone is altered in the liver, how much of the real progesterone are you getting? Is your liver functioning at a hundred percent efficiency?" I didn't have the answers, so I thought it prudent to begin dissolving my pills sublingually.*

At the same time I am quite happy with the results of using the transdermal progesterone cream on a routine basis. I have found that it also has many other uses. For instance, I sometimes massage it where I have back, hip, or knee pains. Doctors now report that rubbing progesterone cream or oil directly onto the joint or painful area helps their patients.[8]

It's important for all of us to ask lots of questions prior to choosing the

---

* Since individual needs can vary greatly, depending on the health of one's liver and adrenals, some may want to obtain an adrenal stress and hormone balance evaluation.

type of product (creams, pills, or drops) to use. We need to evaluate these according to strength, purity, and quality of the delivery system and relate these to the degree of our symptoms and estrogen dominance. I have concluded that it's well worth the time and effort required to move toward the goal of satisfying one's unique requirements.

I've learned a great deal about progesterone, especially since coming upon one of the most informative books of all on the subject: *Natural Progesterone: The Multiple Roles of a Remarkable Hormone.* Its author, John R. Lee, M.D., instructs us not only about progesterone's molecular structure and the interplay of our natural hormones but also about the many advantages of using the cream. "How long should a woman stay on this natural progesterone cream?" Dr. Lee is often asked. He replies, "I want them to stay on it till they are ninety-six and then we'll reevaluate!"[9]

Later, after putting into practice what I knew at last to be essential to my overall health, I learned of the wide-ranging implications of natural hormone replacement therapy for other health problems. Principal among these are cancer, osteoporosis, and heart disease. It is well known that the use of synthetic hormones increases a woman's risk of breast and endometrial cancer; the use of natural progesterone changes those odds. As for osteoporosis, only recently has the importance of this type of hormone therapy become known for treating bone loss. I will show you evidence that natural progesterone therapy can halt and even reverse the effects of osteoporosis. Sections of this book cover each of these subjects, as well as the connection of progesterone to cardiovascular problems.[10]

As we unearth solutions that guide us toward optimal health, we often discover that the answers are found not in just one area of the natural sciences but in many. Raw foods, natural nutritional supplements, and exercise are among the many healthy ways we can encourage the body's innate healing power. They can help us attain both greater mental clarity and a more serene confidence in God's power. Through these channels we can turn our negative stress and anxiety into positive opportunities for growth, learning, and service to others.

VICTORY AND RESPONSIBILITY

The thought of the legions of women who have met conflict and contention with their physicians in the search for better health prompts me to recall the struggle of Helen Keller, as depicted in a recent book about her life, *Light in My Darkness.* I could not help but be affected by what her editor described as Helen's "unwavering faith in God's plan, as she fought and then found through her religion that every human life is of sacred importance and dignity."[11]

Faced with misfortune, we either succumb to our impediments and prejudices, or overcome them. And as Helen aptly expresses it, "Life is either a daring adventure or it is nothing."[12] As we forge ahead with the knowledge we have at hand, we are better prepared for tomorrow's challenges—a bit wiser concerning our choices and more mindful of how to protect ourselves and our dignity.

As for myself, the moment I learned what is good and natural for homeostasis—our internal balance—is a moment of truth I will never forget. I knew then (and my experience confirmed) what I must also convey to others who are suffering from the grim side effects of synthetic hormones and other related medication.

Trust your own intuition. If at first your doctor says, "Oh, you're too young to worry about menopause and hormone replacement therapy," you can ask for tests such as the LH and FSH which measure hormone messages between your pituitary and ovaries. Do this before you are given a diuretic or pills for pain, insomnia, "nerves," or high blood pressure. However, note that although FSH does go up at menopause, Sandra Coney in *The Menopause Industry: How the Medical Establishment Exploits Women,* says there is disagreement among doctors as to the level that confirms the reaching of menopause. Furthermore, she says, "hormones can fluctuate wildly during the menopausal transition (40–50 years old) and even for some months after the last period; . . . therefore, biochemical tests cannot accurately predict whether a woman is menopausal [and] are really not much use."[13]

According to Dr. John R. Lee, Harvard University's Dr. Peter Ellison "has shown that you can get a more accurate measure of the functional level of estrogen and progesterone by measuring it in saliva" rather than in the blood. "It's logical," he says, "and less expensive."[14] To find out where you can order a hormone level saliva test kit to use in the privacy of your own home, see chapter 7. Certain laboratories also perform a more comprehensive version of this test, if ordered by your doctor.

Without any testing, the doctor may just prescribe stopgap medication, with the reassuring words, "This will calm you down so you can get through the day. It will also help you sleep."

Does this sound familiar? And keep in mind that, with so many other patients needing attention, once you have left the office you may be "out of sight, out of mind." In the end, *you alone* are responsible for yourself. No one else is likely to be willing to invest as much time or effort as you are in your own welfare.

I invite you to learn how and why plant-derived progesterone can help prevent the symptoms of PMS, menopause, osteoporosis,[15] fibrocystic breast disease,[16] and painful endometriosis[17] and may reverse disorders ranging from blood clotting to vaginal atrophy[18] and even some forms of cancer.[19] And should progesterone alone not prove effective, I also encourage you to take a look at how some patients and their doctors are considering a combination of progesterone and estriol (or estrogenic herbs) for the treatment of these same conditions (see chapters 2 and 5 for more details).

I hope that for others who, like me, have experienced PMS or childbirth difficulties or the complicated trek through the premenopausal, perimenopausal, and postmenopausal passage, this book will provide some well-deserved solutions.

CHAPTER 2

PROGESTERONE DEFICIENCY, YES; ESTROGEN, MAYBE . . .

The medical mindset of estrogen prescribed alone represents
a victory of advertising over science.

John R. Lee, M.D.[1]

Most women don't want to run to their doctors for every ache or minor illness that comes their way. They'd rather just wait it out and let the body heal itself. Or maybe, like me, they're simply afraid to go—anxious about being misdiagnosed or having to bring home yet another package of potent prescription drugs, with their usual side effects. Or concerned about having to endure a deluge of tests, perhaps to be told that there is nothing wrong and to be presented with a huge bill.

As we age, our bodies can come to seem less friendly, putting us through many tests and challenges. I've seen this particularly in my older relatives and friends as they've begun to suffer from joint pain and arthritis in their fragile bodies. And it's hard to forget my own mother and elderly aunts, maneuvering cautiously and living as best they could with their pain. Nor are they alone among the countless women who have become dependent on family, friends, or nursing homes. Because most have never been exposed to natural hormone replacement therapy (NHRT), they are especially prone to increasing osteoporotic symptoms and heart problems, which force upon them sedentary lifestyles.

The great news is that we can avoid or lessen the severity of these conditions. Our challenge is to overcome our lack of education about what happens in a woman's body at puberty, during the menstrual cycle, and finally just before, during, and after menopause. The average woman is in the dark about the underlying causes of many of the problems she'll encounter at these times. As a result, she may accept treatment or therapy that is not really in her best interest. Unless we all take charge of our own bodies, we will continue to experience the same traumatic consequences.

In the pages to come, we'll discuss all this in detail. But first I would like to review some relevant basics about a woman's hormonal system, define some terms we'll be using, and look at some standard medical therapies.

THE RIGHT BALANCE OF HORMONES

Both progesterone and estrogen are vital to the life and well-being of every woman. These hormones are produced primarily in the ovaries, beginning at puberty and continuing, in the case of estrogen, for the rest of her life. The two hormones exist in a delicate balance, and variations in that balance can have a dramatic effect on one's health. Additionally, the amount of these hormones that the body produces from month to month and year to year can vary, depending on a whole host of factors such as stress, nutrition, and exercise.

Finally, at the onset of menopause there is a radical change: the production of estrogen decreases significantly and the production of progesterone virtually stops. This causes a major shift in the fine balance between the two hormones that the body has attempted to maintain to that point. This imbalance leads, inevitably, to the unpleasant menopausal symptoms many women experience.

This might be a good place to point out that the chemical building block for many of the body's hormones is cholesterol. Not enough emphasis is placed on the importance of our good cholesterol. Cholesterol is the first step in a complex process. It is converted into pregnenolone, which is the precursor of both progesterone and dehydroepiandrosterone (DHEA). From one or the other of these hormones, in turn, come androstenedione, testosterone, and the estrogens. So progesterone, estrogen, testosterone, and DHEA are all made from cholesterol.[2]

Estrogen is thought of as *the* female sex hormone. It is responsible for triggering all of the changes that take place in a girl's body at puberty and for sustaining them in later life, and it plays a vital role in the menstrual cycle. Unlike progesterone, which is a single hormone, estrogen is actually the general name for a group of perhaps twenty different female hormones of very similar structure and function. The most important of these are *estrone, estradiol,* and *estriol.*

In this book we will generally follow the layperson's convention of speaking of estrogen as if it were a single hormone. But bear in mind when we do so that we are really speaking of the actions of one or more of the particular estrogens. We will identify specific estrogens only when such identification is important to the discussion.

Both progesterone and estrogen have many functions in the mature woman's body, but the most important for our initial consideration is probably their role in the control of the menstrual cycle. Here the two opposing hormones work in careful balance to control the woman's reproductive functions, time the cycle, and sustain any eventual pregnancy. On cue from follicle-stimulating hormone (FSH), which triggers an egg to mature, estrogen starts the endometrial buildup and controls the first part of the cycle. Without an adequate level of estrogen, the cycle will not start. Estrogen production gradually builds to a peak just before ovulation, then levels off for the remainder of the cycle, dropping again at the end.

On the other hand, the ovaries dramatically increase their output of progesterone at the time of ovulation, about twelve or thirteen days into the cycle. They are prompted to do so by luteinizing hormone (LH) from the pituitary gland—the same hormone that stimulates the release of the egg. The level of progesterone rises rapidly to a peak in three or four days, surpassing the level of estrogen, and remains elevated in order to develop and maintain the endometrium (uterine lining) in the event of conception. Progesterone dominates and controls the cycle during this latter half.

Progesterone is essential to the survival of the fetus and its continuing development until birth. Its name, in fact, is derived from this principal function: "pro-gestation." It has many other roles in a woman's body, though, and exerts a much broader impact on her health and vitality than might be supposed.

If conception does not occur within ten or twelve days, the levels of both progesterone and estrogen drop quickly, menses takes place, and the cycle starts over again. The dramatic drop of progesterone is the trigger that causes the body to shed the endometrium. In addition, it has been discovered that the presence of sufficient progesterone prior to ovulation (that is, prior to its normal surge) prevents the release of an egg by either ovary. Knowledge of this phenomenon led to the development of birth control pills, which employ synthetic progesterone-like compounds that simulate some of the functions of progesterone.

At menopause the body reduces the production of estrogen and halts the production of progesterone. The amount of estrogen in the body drops below what is necessary to start another menstrual cycle—so no cycle can start. However, low levels of estrogen are still present. The level of progesterone, on the other hand, drops to near zero. For all practical purposes, the female body ceases to produce progesterone. This, then, is the condition of a woman's body after menopause—the presence of reduced amounts of estrogen accompanied by a virtual absence of progesterone.

In optimal circumstances before menopause, any negative physiological effects of estrogen are suppressed by the opposing effects of progesterone. In the menopausal woman, though, the imbalance of these hormones causes the deleterious effects of estrogen to surface. They include tendencies to increased body fat, salt and fluid retention, depression and headaches, and increased blood clotting—root causes of the well-known complaints of many menopausal women. As we shall see later, this same imbalance between progesterone and estrogen is frequently a cause of PMS in premenopausal women.

It's obvious, then, that a large number of women could benefit from the use of supplemental progesterone. It is very effective in treating or preventing the above conditions as well as menstrual irregularity, cramping, miscarriages, infertility, incontinence, endometriosis, hot flashes, night sweats, vaginal dryness, cardiovascular disorders, and more—because it restores the balance between estrogen and progesterone.[3-5]

Until recently the market was waiting for an abundant and inexpensive source of supply. A source was actually identified more than fifty years ago when scientists found progesterone-like substances occurring naturally in numerous plants. One of these in particular was a substance found in the wild yam, called diosgenin.* More important, researchers soon discovered that this natural substance could be easily converted to a compound that is very similar to the body's own progesterone. More recently this has been formulated into a reliable, inexpensive, natural cream.[6] So the supply is now at hand.

THE STANDARD TREATMENT

Unfortunately, natural substances themselves are not patentable and don't yield the large profit margins of proprietary drugs. So the pharmaceutical industry immediately went to work inventing synthetic (and profitable) prescriptive progesterone-like products derived from this very same source of natural progesterone. Soon a whole new class of substances, called progestins, was created by the industry. These progestins are widely used today in birth control pills and are the drugs of choice of the medical establishment to treat PMS and menopausal symptoms. One of them, Provera, was the drug originally prescribed for my own use. Very little attention was given to the possible use of the natural progesterone itself.

---

* Diosgenin is one of a hundred phytogenins (in *Dioscorea* and other plants) that have a hormone-mimicking character.

Just a note about names at this point, because there is a fair amount of confusion about what to call these synthetic substances. Some of the literature, and perhaps some doctors, may refer to progestins by the name "progestogens," "progestations," "progestens," or, in Europe, "gestagens." For the sake of clarity we will use the name "progestin" in this book. Just be aware if you see any of the other names that they are referring to exactly the same group of substances.

Many doctors persist in referring to these substances as "progesterone," but that is a serious error. There is only one progesterone—the natural substance. The progestins so widely prescribed are synthetic substances derived from minute amounts of botanical source materials. They mimic some of progesterone's functions, and they have some advantages for particular applications, but all of them are chemically different from progesterone. This chemical difference causes all of the progestins to have significant side effects, whereas progesterone has no known side effects. That's important enough to repeat: the progestins all have possible serious side effects; natural progesterone produces no known adverse reactions. We'll talk more about this later.

Progestins have some limited advantages. They are administered orally, which makes them convenient to take. Further, the oral tablets facilitate specific and consistent dosage, making them convenient to prescribe. Last, their synthetic nature makes them much more difficult for the body to metabolize (compared with natural progesterone), so they stay in the body longer and have a longer-lasting effect. That's both good news and bad news. The good effects persist, but so do the unwanted ones.

This question of side effects is not to be taken lightly. In his book *Natural Progesterone,* Dr. Lee reprints just an abbreviated list of the side effects of Provera, the most common progestin used in menopausal treatment. He lists five specific warnings (including sudden or partial loss of vision and pulmonary embolism), eight contraindications (existing conditions you may have that preclude use of the drug), ten possible adverse reactions (such as breast tenderness, acne, and weight changes), and five other consequences (such as headache, loss of hair, change in appetite) that have been observed when Provera is taken with estrogens. In contrast, natural progesterone has—I repeat—no known side effects.

THE PERILS OF BEING A PATIENT

Many of us have strong concerns about taking artificial substances anyway. Deep down we know that most of the synthetic hormones and drugs we use won't correct the actual cause of our symptoms but will only temporarily relieve them and camouflage the problem. Yet, we often don't know where to turn. We listen to the doctor as he firmly advises us to continue with the prescribed medication. He gives us a fleeting sense of hope that we are going to put an end to this misery by saying, "You haven't given the medicine enough time," or, "Give it three to six weeks (or months) more."

You may have already experienced some side effects or felt worse since beginning the drug. You wonder, "What do I do now? How can I cope?" In my own case, I tried to find a way to deal with this dilemma and my resentment of a system that was not getting to the cause of my problems. To use the words of author Peter S. Rhodes, I tried "internal considering."[7]

Putting myself in the shoes of the medical doctors who were doing their best to treat me at the time, I said to myself: "Doctors are busy people with full waiting rooms and many interruptions for emergencies. When they go home, they're tired, and it's all they can do to try to keep up with all the latest findings." They must continually deal with new information and the stream of articles in scientific journals. Of course, that heavy schedule also includes business seminars given by drug companies and conferences with their representatives.

Soon, my "internal considering" began to fade. I had tried to give my doctors the benefit of the doubt, but I was well aware of what is certainly no secret: that doctors are heavily influenced by the pharmaceutical companies' sales force and by promotions for various drugs. I realized that the advice they would be giving me could be prejudiced.

Then and there, I decided to substitute action for consideration. The frustration I'd encountered forced me to face my own ambivalence. If I wanted unbiased answers to my questions, the time had come for me to do my own trail-blazing and to take responsibility for my own health. It was time to go to work. The following is what I found.

A VERY PROFITABLE BUSINESS

The puberty-to-postmenopause population provides the drug companies a booming business. This multibillion-dollar bonanza for the U.S. pharmaceutical industry, however, is taking its toll on more than half a million women, especially those going through the midlife crisis. It's no wonder more and more books are being written about the medical establishment's exploitation of women and our need to protect ourselves from becoming "hormonal guinea pigs."[8]

The commercially adulterated substances that are used today to create estrogen and progestin products, notwithstanding their poor utilization by the body, are misrepresented to the consumer as the "fountain of youth." At the same time, the plant-based phytohormones that the body is able to utilize go virtually ignored, having little place in the profit-driven world of drug promotion. What is sad is that misinformation about estrogen is being given to vulnerable women who are most desperately looking for relief.

No matter what "the authorities" say, "Estrogen is a potentially danger-ous drug with significant side effects," warns Dr. Lawrence Riggs of the Mayo Clinic.[9] Nevertheless, the pharmaceutical industry has cultivated a great market among menopausal women by publicizing estrogen as "essen-tial to a woman's good health and her womanhood."[10] However, as sales have been increasing, so have breast and endometrial cancer.

PROMOTING ESTROGEN: A POWER STRUGGLE

Estrogen is considered one of our most potent prescription drugs. In *The Menopause Industry: How the Medical Establishment Exploits Women*, Sandra Coney recounts this sordid tale: "Warnings about the dangers of estrogen had been made sporadically for nearly 30 years. In particular, it was known that estrone, the form of estrogen in Premarin, could be associated with the development of endometrial cancer. As early as 1947," she discloses, Dr. Saul Gusberg of Columbia University "called the ready use of estrogen 'promiscu-ous' and warned that what was going on was a human experiment." He had observed too many estrogen users coming in for dilation and curettage (D&C) for abnormal bleeding caused by endometrial overstimulation, as well as documented cancerous and precancerous changes of the uterus.[11]

With time and with more investigation of the serious problems that were occurring, the FDA finally insisted that all prescriptions be accompa-nied by warnings about the risk of cancer, blood clots, gallbladder disease, and other complications. When this estrogen scare reached the public, sales began to decline. Without a moment to lose, however, the American Pharmaceutical Manufacturers' Association and the public relations firm for Ayerst Pharmaceutical, Hill and Knowlton, wasted no time in produc-ing sales strategies and an intense promotional campaign. This included articles sent out to magazines (*Reader's Digest, McCall's, Ladies' Home Journal, Redbook*) and 4,500 suburban newspapers in order to "preserve the identity of estrogen replacement therapy as effective, safe treatment for symptoms of the menopause."[12]

Those with monied interests were so opposed to the FDA's plan for packaging the warning inserts that they took legal action, for "patient information would reduce sales of estrogen drugs and, therefore, reduce profits." Other organizations that joined in opposition were the American College of Obstetrics and Gynecology, the American College of Internal Medicine, and the American Cancer Society. They claimed that "giving patients information violated the physician's right to control how much information to disclose to patients and threatened medicine's professional autonomy." Eventually the U.S. National Women's Health Network introduced a brief to the court in favor of the FDA,[13] and the FDA won out.

Synthetic HRT remains a booming business. But because of the risks involved with the drugs, a variety of profitable tests, procedures, and drugs is called for along the way—from blood tests, biopsies, and mammograms to hysterectomies, D&Cs, pain relievers, blood pressure medication, diuretics, and frequent doctor visits. And there are as many as 175 different possible treatment combinations to experiment with when bad reactions occur![14]

Many of us don't recognize when we may have been given the wrong type of hormone. We are completely dependent upon what our medical doctors advise with respect to trying new formulations that have just come on the market. Yet, year after year, as we fail to get results from medicines developed through costly technology, we can't help but perceive the undercurrents of greed associated with products that are promoted at the sacrifice of public health. Slowly but surely, we are beginning to think twice about what's in store for us and are asking more questions that challenge the physician's monopoly of information.

What the medical establishment calls breakthroughs are often justified in the name of "consumer protection." However, we need to be mindful and learn as much as we can about what is best for our welfare. I think we should seriously consider the words of John Lee, M.D., over a decade ago concerning this frustrating situation:

> The emerging realization that estrogen should never be given unopposed, i.e., without progesterone, due to its risk of developing endometrial carcinoma makes natural progesterone a valuable addition in those cases where menopausal symptoms require treatment. . . . It is amazing to me that, given the extensive supporting medical references presently existing, estrogen without concomitant progesterone is still commonly prescribed.[15]

Previously I have discussed the different forms of estrogen. In this section I must reemphasize that two forms of synthetic estrogen, estradiol and estrone, are often prescribed in spite of being potentially tumor-forming. "It is believed but not proven that estrone is even more carcinogenic than estradiol,"[16] says Dr. Lita Lee. This is not good news, because orally administered estradiol is mainly converted to estrone in the small bowel.[17]

The next time your doctor prescribes these chemical compounds, you might want to ask this question: Why is Premarin, which consists of estrone and estradiol,[18] being prescribed when these hormones have been implicated as potential causes of cancer?

A study reported by Graham A. Colditz, M.D., in *Cancer Causes and Control* showed a 59 percent increase in breast cancer risk for women who had used synthetic HRT for over five years, and an additional 35 percent risk for those fifty-five years of age and older.[19] In conjunction with Harvard Medical School, Dr. Colditz extended his study of 121,700 nurses for a total of sixteen years. A later report published in the *New England Journal of Medicine* (June 15, 1995) concluded with similar figures and noted "a clear, significant increase in risk" associated with standard, long-term synthetic hormone replacement therapy.[20]

I have personally witnessed more than one of my friends and colleagues, after twenty or thirty years on various estrogenic substances, undergoing a mastectomy of one breast and another mastectomy two years later, and within about four more years they had died. They were denied a decent quality of life in what should have been their golden years.

We must all become more informed so as to avoid a possibly tragic outcome. If you need estrogen for any reason, ask for what may prove to be a safer form, *estriol*, which we'll elaborate on later in this chapter and in chapter 5. Even then, try it only after you have investigated other avenues.

REAL PROGESTERONE—IT'S A NATURAL

The need for natural progesterone is confirmed and reiterated in numerous research papers. Progesterone has been prescribed for more than thirty years with no reported increase in cancer incidence.[21] In fact, a "Women's Health Report" in *McCall's* tells of research indicating that "progesterone deficiency—which women with PMS have—actually increases the risk of developing breast cancer."[22] This article records an astute observation by Phil Alberts, M.D., who heads a PMS treatment center in Portland, Oregon, that the stress that occurs during PMS often triggers ailments that do not seem at all related to one's hormones.

Who would guess that colds, flu, asthma, allergies, epilepsy, migraine headaches, and various endocrine disorders might be connected with a severe progesterone deficiency? Dr. Alberts explains that problems such as these, seemingly unrelated to PMS or menopause, tend to manifest themselves at times when a woman's immune system is depressed.[23] Progesterone is the real missing ingredient for increasing vitality, enhancing sexual libido, and reducing sleep disturbances.[24]

Upon finishing my personal research, I was overcome with strong feelings about the injustices inflicted upon the thousands of women who need this information and desperately deserve to be helped. However, turning my thoughts to a more positive sort of reflection, I began to think of all the medical doctors who are looking for better and more natural ways to help women avoid PMS and menopausal symptoms.

The physicians mentioned throughout this book are among the many who are now making available information on the benefits of natural HRT. More and more, you may come across published accounts by such individuals. For example, Niels H. Lauersen, M.D., says, "In my practice, hundreds of women who were severely handicapped by PMS have been completely symptom-free with progesterone."[25] We can place further reliance on the reinforcement of progesterone when we read in Dr. John Lee's book that progesterone also seems to assume a preventive role in PMS and other conditions.[26]

No wonder we feel gratitude for those who have introduced us to this natural treatment. We need to hear about the new findings over and over again. Otherwise, over and over again we will be enticed into trying synthetic hormones that only steer us further from homeostasis, the hormonal and metabolic balance we want to achieve.

THE PROTECTION AND POTENTIAL OF PROGESTERONE

One of the first things women ask is whether there are any adverse side effects with natural progesterone. All of my investigation says there have been no negative results—only positive. In fact, Dr. Niels Lauersen tells us: "Progesterone is not believed to be cancer causing. No human cancer has been reported during progesterone treatment; quite the reverse, progesterone has been used in treating specific uterine cancers."[27]

Dr. John Lee mentions that when the proper progesterone dose is determined, "because of the great safety of natural progesterone, considerable latitude is allowed."[28] Occasionally, a slight feeling of drowsiness may indicate that you're using more than your body needs.

Not only does natural progesterone have no serious side effects, but it is a precursor of other hormones including adrenal corticosteroids, estrogen, and testosterone. Dr. Lee informs us that it participates in the ultimate formation of all the other steroids and hormones.[29] Progesterone is beneficial in treating or preventing

- irregular menstrual flow; cramping
- bloating; depression; irritability
- migraine headaches; insomnia; epilepsy
- miscarriages; infertility; incontinence; endometriosis
- hot flashes; night sweats; vaginal dryness
- hypoglycemia; chronic fatigue syndrome; yeast infections
- heart palpitations and other cardiovascular disorders
- osteoporosis (reversible by increasing bone mass)[30–33]

With progesterone, blood pressure often returns to normal,[34] body fat is burned up for energy, and cell membrane function is safeguarded.[35] Progesterone not only has an anti-inflammatory effect but also helps balance the cellular fluid, which protects against hypertension.[36]

However, be especially careful if you are taking estrogen for the purpose of preventing heart disease. Epidemiologic investigations and many other studies show that estrogen has no coronary benefit and that its use increases not only the risk of cardiovascular disease but also the risk of stroke or even bleeding from a brain artery. Doctors have been giving estrogen on the basis of a study "limited to postmenopausal women free of any history of cardiovascular disease or cancer." Statistics can be easily manipulated, with mainstream medicine's focus on estrogen and its eagerness to prescribe estrogen hormone replacement therapy. As Dr. Lee says, the media have perpetuated the estrogen myth, even though the hype was built on flimsy evidence.[37]

It should be noted that the severity of endocrine or reproductive system disorders can be affected not only by hormonal imbalance but also by poor diet[38, 39] or by nerve interference within the neuromusculoskeletal system.[40] Many doctors do not adequately address these factors or the probable underlying progesterone deficiency, which is often accompanied by an overabundance of estrogen. Instead, they rely on treatment with antidepressant drugs, aspirin, ibuprofen, other analgesics, or sleeping pills. Fortunately, a much more effective remedy is available in the form of natural progesterone cream, which offers these further benefits:

- protects against fibrocyst formation, especially in the breast
- keeps the uterine lining healthy, helping to prevent fibroids, etc.
- assists thyroid hormone action

• normalizes the blood-clotting mechanism
• restores libido (sex drive)
• acts as a natural antidepressant[41-43]

Adverse symptoms can begin when a woman is in her thirties or even at the onset of menses in her teen or preteen years. So it is important to be thinking about natural alternatives in the years prior to menopause. Prevention is essential for any health condition, and the sooner we look into natural sources, the sooner we can start to think, look, and feel as young as we are. After studying what natural progesterone does for the bones, the heart, and the body as a whole, we can better understand the need for it as part of a natural HRT program.

Now You Don't Have to Refuse Hormone Supplements

I so often hear women repeating the same thoughts I once had: "I won't take hormones. I don't believe in taking pills." That's because many people don't make the distinction between natural ingredients and most of the drugs that are continually portrayed on TV. Yes, NHRT is in a different class, as it represents a natural replacement that the body needs; and yes, we do need to continue to resist commercial inducements to take pills, and to try to ignore the medical hype.

Concerning the need for NHRT, Dr. Betty Kamen states in her book *Hormone Replacement Therapy: Yes or No?* that even if you don't have menopausal symptoms and have a good diet and exercise regularly, the use of natural progesterone is still recommended to fortify one's body for today's steady diet of stress. And to be realistic, she points out, no diet is perfect anyway; and we all cheat on top of that.[44] No matter how diligent we are in maintaining a proper diet, there are always days when we want to escape from the stresses of life, saying, "Everything will be OK. Here, have a treat—something sweet. It will make you feel better!"

Cravings for ice cream, doughnuts, or other simple carbohydrates can be pretty powerful. When stress takes over your senses, no one has perfect control, and if you do, you are the exception. It's easy to forget in that moment that the sugar will only stress one's system more. But, while using natural progesterone does not justify such lapses, we can take some comfort in the knowledge that it helps support the adrenals and our stress glands and helps protect against hypoglycemia.

Dr. Kamen confirms my own experience as to the need for natural progesterone when she says, "Perfect lifestyles/diet may be impossible for perfectly good reasons. Don't feel guilty. Feel better! Natural progesterone could make the difference."[45]

Natural HRT has made me feel like myself again. And I truly believe it will begin to reverse whatever damage the synthetic HRT may have done to my body. Fortunately, I did take antioxidants back then (and still do) to fight any free radical damage, and I added other vitamin and mineral supplements to my daily diet to help counter what might have been toxic to my system.

It is interesting to note that progesterone, a precursor of other hormones, is so nearly perfect for our body chemistry that even its promoters can't exaggerate its importance. Without it I can testify that I felt stressed, run-down, and dependent on medical help; with it, I feel energetic, calm and, most important, free. Menopause does not have to be treated as an illness. It can be better viewed as a challenge! Once a woman establishes proper hormone balance through natural methods, she'll find she's taken a great step towards increased vitality.

It's Safe, It's Sound, It's Easy

For some practical guidance, let's look now at two principal means by which to supply natural progesterone most efficiently to the body: either oral capsules (taken by mouth) or transdermal cream (applied directly to the skin). Some other methods that may be useful will be covered in appendix A.

The most popular way to apply natural progesterone is with the cream. Dr. John Lee reports that he has been using transdermal natural progesterone in postmenopausal women since 1982 and has seen remarkable success.[46] "Progesterone," he says, "like all gonadal steroids, is a relatively small and fat-soluble compound which is efficiently and safely absorbed transdermally. Not to use it in cases of progesterone deficiency is imprudent, to say the least."[47] He cautions, however, that any product containing mineral oil may "prevent the progesterone from being absorbed into the skin."[48] Furthermore, according to Dr. Raymond F. Peat, certain components of mineral oil (which is in many cosmetics) are toxic, and any that does get into the system does not metabolize.[49]

The cream is now available in many brands and formulations and at varying strengths. Researchers have measured the levels of hormone in women who were using various wild yam creams that do not contain USP (U.S. Pharmacopeia) progesterone. While many of these creams worked well, some were found not to have much effect. However, according to Aeron LifeCycles Clinical Laboratory in San Leandro, California [(800)

631-7900], most of the creams that contained USP progesterone did produce hormonal changes in most women.[50] (See chapter 4.)

Christiane Northrup, M.D., in her book *Women's Bodies, Women's Wisdom*, also recommends natural progesterone over synthetic because it is compatible with the body and does not have the side effects (bloating, depression, etc.) produced by progestins.[51] Dr. Peat agrees that transdermally applied progesterone is effective for most symptoms as well as for long-range maintenance.[52] Dr. Lee points out that in the beginning stages of treatment some of the application may be retained in the subcutaneous fat layer sometimes delaying the initial physical response. However, the longer a woman uses the cream, the greater the benefits.[53]

Alleviation of symptoms as a clinical response provides a good yardstick for comparison. As with many forms of health care, starting with the least invasive product (in this case the mildest) gives you a baseline for evaluating your response. You can always switch to a stronger product or add to your initial treatment program.

During the years of menopause, I personally found the use of the transdermal progesterone cream to be adequate for my needs. As I proceeded into postmenopause, however, I felt that the micronized natural progesterone, taken orally in combination with estriol, was more effective in meeting my body's demands. During periods of excessive stress, I add the cream to my program as well, since the progesterone is a precursor of the adrenal hormones.[54] This, in combination with daily nutritional supplements, seems to take care of all my postmenopausal problems.

Micronization Enhances Effectiveness

In the past, progesterone taken orally was considered ineffective because of its poor absorption rate[55] and its premature removal by the liver. Recent advances, however, have changed the picture. Breaking the progesterone down through the process known as micronization has been shown to enhance the rate of absorption and hence the effective level in the body.[56] And according to Rabbi Eric Braverman, M.D., "When compounded in an oil base, the progesterone is so firmly held by the oil base that it is actually absorbed through the lymphatic system first, thereby allowing a couple of passes through the body before being cleared via the liver."[57]

Several pharmacists I have talked to said essentially the same thing. Dr. Peat explains how this works: "If progesterone is perfectly dissolved in oil, it . . . is not immediately exposed to enzymes in the wall of the intestine or in the liver. People often speak of 'avoiding the liver on the first pass,' but in fact chylomicrons [microscopic fat droplets] pass through the liver many times before they are destroyed."[58]

One knowledgeable pharmacist at Women's International Pharmacy, however, told me that if a person's liver is not functioning fully because of compromise by alcohol or disease (such as hepatitis), the cream form may be more effective. He described the transdermal progesterone as "productively potent" because it acts systemically before it acts locally. This process can be quite beneficial, as it cuts down on some of the work the liver has to do.

Progesterone is not the only hormone that can be micronized. We have referred previously to the estradiol form of estrogen as being harmful. However, for those readers who are reluctant to give it up, please note that studies in which estradiol has been broken down, through micronization, for utilization by the body have ascertained that when natural progesterone is prescribed along with minute doses of the micronized estradiol, both hormones together are at least safer than the standard synthetic HRT. (Nevertheless, if estrogen is deemed absolutely necessary for long-term use, estriol should be considered, and will be discussed in the next section.)

First, let's look at one trial involving a "combination of micronized estradiol (E2) (0.7–1.05 mg) and natural progesterone (200–300 mg) given to ten menopausal women with vasomotor problems and/or vaginal atrophy. Five other women," the report says, "were placed on a daily course of conjugated estrogens (0.625 mg) with medroxyprogesterone acetate (10 mg)."[59, 60] All women in the first group (on E2 and nonsynthetic progesterone) had a decrease in total cholesterol and an increase in high-density lipoprotein cholesterol. Those on conjugated estrogens and medroxyprogesterone acetate (Provera), on the other hand, had no meaningful changes in their total cholesterol but some increase in their HDL. In this study they found: "Most significantly, the adverse effects of synthetic progestins on lipoproteins and cholesterol were eliminated by using natural progesterone."[61]

The experiment showed that "administration of micronized E2 and progesterone results in symptomatic improvement, minimal side effects . . . improved lipid profit [fat balance], and amenorrhea [no menstrual periods] without endometrial proliferation or hyperplasia [precancerous uterine abnormalities] in menopausal women."[62] *Obstetrics & Gynecology* says in its evaluation, however, that women who still have a uterus and have been on estrogen therapy for a prolonged period of time, without the addition of some form of progesterone, are at increased risk of endometrial hyperplasia and adenocarcinoma:

The problem of endometrial hyperstimulation induced by estrogen therapy is obviated by the addition of cyclic progestin to the hormone replacement. . . . A number of synthetic estrogens and progestins have been used for postmenopausal replacement hormonal therapy, both cyclically and as a daily combination.[63]

The review goes on to say that there are disadvantages to the synthetic compounds.[64] Indeed, study after study confirms our need to choose hormones in their natural form over the synthetic versions. And for women concerned about possible side effects of the estrogens, using progesterone alone would be the most sensible course to follow.

In one set of trials, various preparations of progesterone were used: (1) plain-milled, (2) micronized (more finely ground), (3) plain-milled in oil, (4) micronized in oil, and (5) micronized in enteric-coated capsules. All of these preparations were administered orally to six postmenopausal women in order to establish which combination would achieve the best absorption rate. These were the findings:

Micronized progesterone in oil showed the highest average progesterone concentration. . . . [A] decrease in the particle size of progesterone through micronization increased aqueous dissolution in the intestine and further enhanced absorption. . . . Progesterone taken orally is physiologically active, producing a significant increase in tissue progesterone concentrations in breast, endometrium, and myometrium.

On the basis of the results of this study, the optimal preparation for the administration of natural progesterone should include micronization of the progesterone particles and dissolution in oils consisting of principally long-chain fatty acids.[65]

Other important research exists, however, showing much greater effectiveness and less toxicity to the tissues with a base of natural vitamin E (tocopherol—*not* the cheaper tocopherol acetate some manufacturers use).[66] In the same journal it is stated that "oral progesterone has not shown any adverse effect on the beneficial changes in serum lipoproteins induced by the administration of estrogens."[67] In fact, a major medical journal consultant has commented that the concept of dissolving progesterone in vitamin E for absorption into the lymphatic system "is so simple it is amazing that the pharmaceutical companies have not jumped on it."[68]

The aforementioned study, along with thirty-three others in the report, demonstrates that a micronized natural progesterone preparation given orally raises blood levels of the hormone with no observable adverse side effects (such as bloating, breast tenderness, weight gain, and depression), all of which are induced by synthetic hormones.

The *American Journal of Obstetrics and Gynecology*, while still advocating estrogen therapy, concedes: "Oral natural E2 and progesterone would seem preferable for long-term replacement therapy."[69, 70]

THE OVERLOOKED ESTROGEN

You've probably heard a lot of discussion these days about estrogen. It seems that most women of a certain age are either on it, pondering whether to start it, or contemplating getting off it. Some have been made to believe that it is the cure-all for PMS and menopausal discomfort, heralded to end all our female problems. In chapter 3 we will find that this type of promotion has led to widespread suffering and disease. Here we will review another estrogen now attracting interest: estriol, which is not normally prescribed by your doctor.

In its pure tablet form, estriol has been administered in thirty different countries for more than thirty years.[71] One of the advantages it has over other estrogens is its use for urinary or vaginal symptoms, with little or no contraindication to the uterus. Estradiol (brand names: Estrace, Estraderm) and estrone (brand names: Premarin, Ogen) cause the lining of the uterus to thicken, thus increasing a woman's risk of endometrial cancer (further discussed in chapter 5). Estriol, on the other hand, has a weaker stimulatory effect on the uterus.[72]

In any form, unopposed estrogen supplementation as usually prescribed may put you at risk unless botanical progesterone is used to counter any excess of estrogen. So if and when you contemplate estrogen therapy, consider the possibility of combining progesterone with estriol or estrogenic herbs (see chapter 6). Some companies, recognizing the danger of using estrogen alone, are now marketing an estrogen/progestin combination pill—but as we know, a synthetic version is not the same.

Estriol for Managing Infections and Restoring Tissue

A couple of studies have found that with the use of estriol, vaginal tissue is restored to a normal, healthy condition.[73] With as little as 3 mg daily of estriol, vaginal flora were renewed.[74] Furthermore, vaginal application of the unconjugated estriol cream (not synthetically compounded) was found to be more effective than estriol taken orally at the same dosage.[75]

Beneficial results were also found in a controlled study of postmeno-pausal women with a history of chronic urinary infections. The reduction in urinary infections was significantly greater with those on estriol than with those on a placebo.[76] In another study, normal pH was restored after 1 mg of oral estriol was administered daily for one week.[77]

The clinical studies found no contraindications when the proper doses of estriol were prescribed.[78] No water retention was observed in the clinical studies,[79] nor changes in blood pressure[80] or abnormalities in routine uri-nalysis or cholesterol levels.[81–83] By contrast, the forms of hormones that are usually prescribed, which are compounded with synthetic substances, have indeed been shown to result in the aforementioned conditions.

One downside to estriol is that there have been no studies concerning its effect on bone mineral density. It is often presumed to be the weaker estrogen (although some studies have found the contrary)[84] and thus is also weaker than estradiol and estrone in retarding bone resorption. During pregnancy, estriol levels are extra high at a time when the body may be drawing on calcium and other minerals from the mother's bones to supply the developing skeleton of the baby. Based on this reasoning, if a woman is not at high risk for cancer, Dr. Lee would prescribe progesterone as the primary hormone, but might also add some estriol in combination with estradiol and estrone, a ratio of 4:1:1 or 1:1:1. The latest research finds that estradiol is transformed into estrone in the intestinal tract. Because of the strong implication of estrone in hormone-mediated cancers, some physi-cians prescribe bi-estrogens (80 percent estriol and 20 percent estradiol) along with the progesterone.[85]

Should I Use Estriol?

What might be completely effective for one woman may not be the complete cure-all for another. If you find that hot flashes and other discomforts are still bothering you even after using natural progesterone (as directed and for a minimum of three months), you might try a stronger source of yam extract cream, a USP progesterone, or a prescriptive product before considering supplemental estriol. Although I chose to do the latter, those who know they are at high risk for breast cancer should think twice about using any estrogen—or an excess of phytoestrogens—until we have more knowledge of their consequences. In clinical investigations, hot flashes were diminished when estriol was used in doses of 1 to 8 mg daily, depending on the patient's need.[86, 87] In most patients, estriol resolved problems with not only vaginal atrophy, but also headaches, insomnia,[88] irritability,[89] nervousness,[90] tiredness, heart palpitations, and depression.[91]

It also helps with urinary incontinence, which often affects postmeno-pausal women because, according to a scientific abstract, "in the postmeno-pausal [woman], the urethra becomes narrower and more sensitive to the passing of urine."[92] Progesterone alone may help some women with incontinence, as a deficiency of this hormone, according to Dr. Ray Peat, makes the bladder more sensitive. Others have felt the need to try estriol. In one study, improvement was considerable when patients were treated with 3 mg of estriol daily.[93]

Dr. Alan Gaby contends that the use of estriol could reduce the need for D&Cs and even unnecessary hysterectomies, because it rarely produces endometrial bleeding. Estriol is also effective in lowering the risk of blood clots in the lungs and veins. Dr. Gaby comments in his book *Preventing and Reversing Osteoporosis* that to relieve menopausal symptoms, "a dose of 2 to 4 mg of estriol is considered equivalent to, and as effective as, 0.6 to 1.25 mg of conjugated estrogens or estrone."[94] Of the three estrogens that naturally occur within your body, some doctors believe that estriol should be dominant. Vitamin E supplements can increase the ratio of estriol to estradiol and estrone.[95]

During my study I read an article written in 1991 by Dr. Lita Lee, who reported that she was not able to obtain estriol and that it was not easily available in the United States. She was limited to using a homeopathic estrogen, of which she said, "It works. . . . No more hot flashes!"[96] Today, however, estriol in its natural form is very much available in the United States.

Dr. Lee points out that except during pregnancy, less than 1 percent of the estrogen we normally make is estriol. However, because of certain preliminary studies regarding cancer and many women's positive experiences with this hormone, he agrees that more study is certainly warranted.

Dr. Marcus Laux prescribes "tri-estrogen" therapy to his patients. For some postmenopausal women he recommends a ratio of one part estrone, one part estradiol, and eight parts estriol, along with application of natural progesterone cream to the breasts.[97] Indeed, in *Fertility and Sterility* (April 1995) we find evidence that natural progesterone applied directly to the breast does offer protection against estrogen's stimulatory effect on breast cells.[98] Other ideal skin sites are the places where the capillaries are abundant and closer to the surface, such as the areas where we blush and the hands and inner arms. Since the skin of the lower abdomen, thigh, and back is thicker and less well supplied with superficial capillaries, absorption at these sites is less efficient.

Excess Estrogen Equals Excess Weight

Contrary to popular opinion, researchers make it clear that menopause does not mean an absolute end to estrogen production. Two other parts of the body besides the ovaries produce estrogen: the adrenal glands and the fat tissue. These, says author Sharon Gleason, "will maintain low levels of estrogens to minimize symptoms."[99]

Dr. John Lee states that a sign of *estrogen dominance* is weight gain caused by both water retention and fat deposition at hips and thighs. This is an interesting point, because I have found that many women wonder why they are gaining weight even though they are exercising and on a strict diet. One of my neighbors thought she had gained weight because she had given up smoking. At the same time, however, her doctor had put her on estrogen therapy for her menopausal problems, informing her (albeit incorrectly) that this hormone would be good for bone growth and make her feel a lot better. After talking to her I couldn't help but think that her accumulation of a middle-age spread could very well stem from the dual effect of her synthetic estrogen (prescribed alone) and the rapid drop in progesterone level that accompanies menopause.

Another case is highlighted in *The Menopause Industry*, by Sandra Coney. This woman, prescribed Premarin for her joint problems and pelvic inflammation, began to put on weight "at an alarming rate" and was then switched to different forms of estrogen, including Estraderm patches and implants. The unfortunate result was a thirty-five-pound weight gain, fluid retention, and breast discomfort.[100]

In contrast, botanical progesterone is a natural diuretic. It burns fat (often caused by high doses of synthetic estrogen already in the body) for energy and lowers cholesterol levels[101]—once again helping to avoid another unwanted side effect of synthetic HRT.

Other Side Effects of Estrogen Dominance

Dr. John Lee proposes further reasons why estrogen should not be given without natural progesterone, and why we may begin to feel more and more uncomfortable if we take estrogen alone for any length of time. He says that estrogen "allows influx of water and sodium into [the] cells, thus affecting aldosterone production leading to water retention and hypertension. Estrogen," he continues, "[also] causes intracellular hypoxia [oxygen deficiency], opposes the action of [the] thyroid, promotes histamine release, promotes blood clotting thus increasing the risk of stroke and embolism, thickens bile . . . promotes gall bladder disease [and] causes copper retention and zinc loss."[102]

Is it any wonder that so many women feel miserable when using synthetic estrogens? And is it any wonder that Dr. Lee says, "Something is wrong with the estrogen theory"? Prescribed alone, estrogen can lead to breast or uterine cancer even five years prior to menopause.[103] Other consequences of estrogen dominance, he says, include "heightened activity of the hypothalamus [and] hyperactivity of adjacent limbic nuclei leading to mood swings, fatigue, feelings of being cold, and inappropriate responses to other stressors."[104]

Just before menstruation, as Dr. Lee says, too much estrogen in the body often causes edema, or swelling and bloating. Dr. Ray Peat agrees: "Under the influence of estrogen, your body retains extra water."[105] This, he says, is one reason we often crave extra salt.

Some authorities recommend cutting down on salt the week before one's period in an effort to reduce bloating and breast tenderness. However, Dr. Peat points out the often overlooked fact that sodium "is essential for [maintaining] adequate blood volume, and that it is almost always unphysiological and irrational to restrict sodium intake." He explains that "reduced blood volume tends to reduce the delivery of oxygen and nutrients to all tissues, leading to many problems."[106] (For more on the wise use of sodium, refer to Jacques de Langre's book *Seasalt's Hidden Powers*.)

FIBROCYSTIC BREASTS

Estrogen dominance in the body causes fibrocystic breasts. However, Dr. Lee assures us, "Restoring hormone balance with natural progesterone usually results in prompt clearing of the problem. . . . When natural progesterone is used . . . during the two weeks before menses, fibrocystic breasts revert to normal within 2–3 months."[107] One patient, who came to him fearful of breast cancer, reported having undergone repeated needle drainage and biopsies. But as one might expect, after a course of progesterone and an improved diet, not only had her cysts disappeared, but many other symptoms were also relieved.[108]

Dr. Lee's instructions for use of the cream are quoted in the popular book *Alternative Medicine*. "Using this progesterone transdermally," he says, "from day fifteen of the monthly cycle to day twenty-five will usually cause breast cysts to disappear."[109]

Concerning fibrocystic breasts, Dr. Nina Sessler says:

Avoiding caffeine and other methyl xanthine derivatives such as black tea, most colas, and chocolate, as well as many nonprescription and prescription medicines which contain methyl xanthines, has been shown to help a great deal with the discomfort. Many physicians recommend vitamin E (400–800 I.U.) and . . . vitamin C can also help reduce the inflammation that often accompanies FBC [fibrocystic breasts].[110]

On the other hand, noted breast surgeon and author Susan M. Love, M.D., states that most studies of caffeine and benign breast disease have been either inconclusive, unscientific, or contradictory and that the popular perception of a connection may or may not be a reality. She points out that individual physiological differences could account for caffeine's affecting one person but not another.[111]

Several years ago, Dr. Linda Force had surgery on her breast to remove a fibrocyst. Following the surgery her breast swelled to twice the normal size and was very painful. The surgeon had not removed all of the cyst because it would have created too much deformity. Dr. Force says, "As time went on, I controlled the problem by watching my diet and avoiding caffeinated beverages. But when I was thirty-five and developed PMS, the breast discomfort intensified. This is when I went to Dr. William Douglass, who dispenses rectal progesterone therapy over a three-month period. For those months only, I was fine. After that, my discomfort was not as bad as it had been previously, and I learned to tolerate it. But every month when my period would start, my breasts would swell and become very sore."

Fifteen years after the first surgery to remove her fibrocysts, Dr. Force was still suffering from the cysts, which were getting more and more painful as time went by. Her medical doctors were advising her to undergo surgery again if they didn't clear up. That's when I provided her with literature about the transdermal progesterone cream. It explained how the decline of progesterone can create estrogen dominance which in turn can cause any number of disorders such as fibrocysts, weight gain, endometriosis, depression, and more. Her first question was, "Is it natural?" I assured her that it is botanically derived and that she had nothing to lose.

As a doctor, she quickly understood the dangers and side effects of unopposed estrogen. The fact is that when estrogen is administered to women with fibrocystic breasts, their condition becomes worse. However, it is easily treated with progesterone therapy.[112]

Beginning her treatment immediately, she conscientiously applied the cream to the sites of the lumps in her breast and to her abdomen twice daily, once in the morning and once at night. As the weeks went by, she saw subtle improvements (more regularity and less clotting) during her periods. After three months, her fibrocysts had disappeared and she was free of the pressure and pain she'd previously had. Her relief at having avoided the prescribed surgery and drugs was evident. And as a true health provider, she soon made this important information about natural progesterone cream available to all her patients and staff.

UTERINE FIBROIDS AND OVARIAN CYSTS

Dr. John Lee refers to fibroids that develop in the uterus as

> another example of estrogen dominance secondary to anovulatory cycles and consequent progesterone deficiency. They generally occur in the 8–10 years before menopause. If sufficient natural progesterone is supple-mented from day 12 to day 26 of the menstrual cycle, further growth of fibroids is usually prevented (and often the fibroids regress).[113]

Ovarian cysts are also a problem in many women. Dr. Peat says these are usually associated with a low thyroid condition, and that administration of thyroid hormone can get rid of them by lowering estrogen levels and making the ovaries produce more progesterone.

Dr. Lee's approach, on the other hand, is to administer just the progesterone directly. He says that "natural progesterone, given from day 5 to day 26 of the menstrual month for two to three cycles, will almost routinely" cause disappearance of these cysts by suppressing normal FSH (follicle-stimulating hormone), LH (luteinizing hormone), and estrogen production and giving the ovary time to heal.[114] Furthermore, studies have been reported in the *Journal of the National Cancer Institute* as far back as 1951 in which progesterone even produced evidence of regression of cervical tumors.[115]

It's reassuring to know that progesterone can protect us in so many ways; but we must all be alert to the fact that the long-range harmful effects of "estrogen dominance" in the body are not widely recognized.

ENDOMETRIOSIS

Majid Ali, M.D., calls endometriosis, which he says afflicts five million American women, "a painful, often disabling disorder that can lead to infertility." It is sometimes treated, mistakenly, with synthetic birth control pills. He blames estrogen "overdrive" for the "growth outside the uterus of misplaced cells that normally line the uterine cavity."[116] Linda G. Rector-Page, N.D., Ph.D., adds that this tissue often attaches to other organs, and there is a backup of some of the heavy menstrual flow.[117]

Dr. Ali maintains that treatment with synthetic estrogens, so widespread among doctors, is a grave error. In fact, *Women on Menopause,* by Anne Dickson and Nikki Henriques, reveals that unopposed estrogen was first linked in 1970 to "abnormal cell growth in the endometrium," resulting also in the possibility of endometrial cancer.[118]

Today, women need to be aware of the many other serious side effects when estrogen is administered alone and their progesterone levels are down: nausea, anorexia, vomiting, headaches, and fluid retention leading to weight gain. It is important, say the authors of this book, for women who have other physical disorders to avoid supplementation with only estrogen, for it can exacerbate high blood pressure, diabetes, migraine, and epilepsy. A study in Sweden also showed that women using high doses of the synthetic estrogen known as ethinylestradiol (used in lower doses in the birth control pill in the United States) had an increased rate of breast cancer.

Sandy MacFarland, who was suffering from endometriosis, was only nineteen when her gynecologist suggested she have a hysterectomy. According to the Endometriosis Association, this condition, which affects girls and women from the ages of eleven to fifty, is "the leading cause of hysterectomy."[119] Fortunately, Sandy's father was a nutritionist, and he decided to try to correct what he thought might be a hormone imbalance with natural progesterone. This decision not only saved Sandy's uterus but also normalized her once irregular periods.

HYSTERECTOMIES: FORCED MENOPAUSE THROUGH SURGERY AND SYNTHETIC HORMONES

Hysterectomies are recommended for numerous reasons, often unwarranted. They are frequently suggested when women complain of adverse reactions to their prescribed estrogen or progestins. A doctor may recommend a hysterectomy instead of offering natural therapies that facilitate the body's own healing process. Gail Sheehy, in her book *The Silent Passage,* tells us that doctors will justify to their patients a more radical surgical approach by explaining that they won't have to take the hormones that have been causing such irritating side effects. The surgery will free them from the worry of having to protect their uterus with hormones.

Gail Sheehy asked one doctor if he took the ovaries out as a routine procedure. The doctor nonchalantly told her, "In a postmenopausal woman, the ovaries are of no use anyway." In dismayed retrospect, Sheehy asks, "Wasn't this extreme?" This doctor was disregarding the fact that the ovaries continue to manufacture testosterone, which, as Sheehy points out, "strongly influences a woman's sexual desire and energy."[120]

We might want to think several times before considering a hysterectomy. Every organ has an integral role to play throughout one's entire life. In her book Sheehy tells us that "between 33 and 46 percent of the women whose ovaries had also been removed complained of reduced sexual responsiveness."[121] Dr. Howard Judd of UCLA, an expert on the postmenopausal ovary, emphasizes that "the concept that the ovary burns out is not true."[122] The fact is that after menopause, even though the ovaries no longer produce estrogen, they do manufacture testosterone.

Sheehy relates that most women who have undergone hysterectomies are in the age group of twenty-five to forty-four. When the cervix and uterus are removed, some women feel the effects of menopause within two years of the surgery. However, an oophorectomy (where the ovaries are removed) will generally bring on the state of menopause immediately. So, coerced menopause often befalls women when they are quite young.[123]

Menopause usually begins around the ages of forty-five to fifty; and the last period is experienced in the early fifties.[124] However, ovarian defect can begin at thirty. Undue trauma, or more than normal physical or mental stress, can bring on menopause years sooner. If a woman smokes, has a poor diet, is on medication, or has undergone surgery, chemotherapy, or radiation, she will also experience a dramatic loss of progesterone, which accelerates the aging process. And the degree of menopausal symptoms can also vary enormously in accord with each individual's genetic differences.[125]

STROKES AND BLOOD CLOTS

My personal experience with blood clots began two months after the laboriously long and difficult delivery of my son. As I was adjusting to my newborn baby, who had colic throughout the night, the meaning of stress became clear beyond doubt. Dr. Lee has mentioned that when stress is heightened, a woman is predisposed to anovulatory cycles (menstrual periods with no ovulation). And conversely, he says, "Lack of progesterone interferes with adrenal corticosteroids by which one normally responds to stress."[126] Also during the immediate postpartum phase, progesterone levels are near zero until ovulation resumes.

I had no way of understanding then the underlying and multiple reasons for my stress and immense weakness, or why my body could not adapt to the demands of taking care of an infant during the night and a two-year-old during the day, while cooking for family and friends who had come to join us at this time of celebration.

As the days proceeded, I rapidly lost motor control on my left side. Within two months after the birth, my left side had become paralyzed as the result of a blood clot that had lodged in a blood vessel wall on the right side of my brain. This embolism left me helpless and traumatized for several months. The neurosurgeon, my OB/GYN, and other specialists were completely mystified as to the cause of my condition.

This postnatal paralysis all took place twenty-six years ago. Now, however, after sifting through much research, I can't help but speculate that progesterone deficiency was perhaps one of the causes for my stress and that natural progesterone might have rescued me from the trauma I had endured. I recall, too, that Dr. Peat writes that during stressful times in a woman's life, supplementation with the hormone progesterone is urgently needed to correct imbalances in the endocrine system.[127] Have these doctors and researchers solved the mystery that was so puzzling to my specialists over two decades ago? Progesterone may help many at-risk women to avoid strokes and other stress-related disorders.[128]

THE HORMONAL DUET

Needless to say, the above experience took place during my childbearing years. Comparing the premenopausal stages of life with the menopausal and even postmenopausal stages, we view hormone replacement in different ways. Our needs differ depending on many factors: our symptoms, age, diet, amount of exercise, level of stress, and other lifestyle habits.

We also need to remember that the menopausal or postmenopausal woman does have some estrogen—it's the progesterone that is no longer produced anywhere in her body to any noticeable degree. It makes one wonder why medical doctors have been prescribing estrogen replacement for so many years—often to the exclusion of progesterone, the disregarded hormone.

Estrogen and progesterone need each other, as each of them sensitizes receptor sites for the other. As Dr. John Lee says, "The presence of estrogen makes body target tissues more sensitive to progesterone and the presence of progesterone does the same for estrogen."[129] It appears that these two natural hormones in proper balance, and *unadulterated*, have a harmonious mutual affiliation.

Dr. Betty Kamen writes, "Estrogen regulates the neurotransmitters of the brain—substances which control the function of our nervous system," including the thinking processes and motor activity. Since this process permits the brain cells to communicate with each other,[130] it may be of value to have one's hormone levels checked when severe PMS or menopausal discomfort arises. Many women can attest to Dr. Kamen's observation that if either hormone level drops, "All hell breaks loose."[131]

On the other hand, according to Dr. Peat, estrogen in excess can act as an "excitotoxin" to the brain, causing an energy drain (cellular exhaustion) by stimulating the brain beyond the nervous sytem's capacity to respond. He says that even an average estrogen level can be a serious problem when there is insufficient progesterone to balance it and that "it is best to have five of ten times as much progesterone as estrogen."[132]

SELF-ASSESSMENT PROFILE

Characteristics of Estrogen Dominance	Characteristics of Progesterone Supplementation
Weight gain	Utilizes fat for energy
Insomnia	Calming effect
Uterine cancer	Stops cells from multiplying
Fibrocystic breasts	Protects against fibrocysts in breast
Breast cancer risk	Helps prevent breast cancer
Depression	Natural antidepressant
Fluid retention (bloating)	Natural diuretic
Thyroid imbalance	Assists thyroid hormone action
Blood clots	Normalizes blood-clotting mechanism
Migraine headaches	Restores oxygen to cells
Risk of miscarriage	Prevents miscarriages
Inflammation	Precursor to cortisone
Cramping	Relieves cramping
Elevated blood pressure	Regulates blood pressure
Acne	Aids in skin disorders
Irregular menstrual flow	Normalizes periods
Restrains bone mineral depletion	Stimulates bone mineral density

Opening Up a New World of Hope and Healing

CHAPTER 3

THE SEASONS OF
A WOMAN'S LIFE

To be surprised, to wonder,
is to begin to understand.
José Ortega y Gasset[1]

There is no reason for an active woman to endure the symptoms of hormone imbalance when safe, natural relief is readily available. We all have enough on our hands trying to solve the host of life's other challenges without adding more. If we don't take the time to provide for our needs, even menial tasks and minor problems can take us over the edge. An incident as simple as having to answer a child's question or give someone directions becomes unbearable. Discomfort, dizziness, or fatigue brought on by menstruation or menopause weakens one's energy and enthusiasm. How easy it is to lose stamina, confidence, and the ability to help oneself, one's family, and others. But simple preventive measures can empower today's woman to rise to any occasion and meet the daily demands of business and family.

The women of this generation need to take heed of what nature has not provided for those of past generations. They did not have full knowledge about hormone therapy, nor did they have the alternative choices we have today. You might have been blessed with one of those "super moms" with genetic vitality and strength. However, many women suffered from heart disease, osteoporosis, or cancer because the slowly diminishing supply of their own natural estrogen and progesterone created an imbalance between these two hormones, and eventual disease.

Now, with the incredible number of artificial hormones that have been introduced (see appendix B), women may encounter unexpected and unpleasant side effects such as bloating, weight gain, emotional tension, and insomnia. And as the years pass, they may have to deal with endometriosis, thyroid disorders, fibrocysts, heart disease, osteoporosis, and cancer. Often labeled chronic complainers, some become dependent on psychiatric care to cope with their pain and to deal with the emotional problems brought on by these health conditions. Let's take a look at some of the beginning signs of hormone deficiency and the symptoms of this cruel onslaught.

YOU'RE NOT CRAZY; IT'S JUST YOUR HORMONES

Long ago, in another generation, women experiencing symptoms of PMS or menopause were often accused of insanity, and even institutionalized. (This is where we get the word *hyster*ectomy.) As early as 1931, several years after both PMS and menopause had been classified as forms of emotional and psychological disturbance, literature on the subject actually admitted that there were physical problems associated with the "change of life."

Today, women are not routinely institutionalized for exhibiting symptoms of hormonal stress, but at the mercy of impulsive moods, hypertension, irritability, depression, and crying spells, we are often diagnosed as having a "nervous breakdown." Many types of medication are prescribed, from sedatives to muscle relaxants. In Great Britain, says Sterling Morgan, "[natural] progesterone treatment for PMS is so accepted that in three different murder trials, women were sentenced to take progesterone. Their defense was that they had committed violent crimes because they were pre-menstrual!"[2]

Dr. Katharina Dalton, in her book *Once a Month*, tells of her experience with patients who exhibited psychological and physical symptoms ranging from what she calls "cyclical" criminal acts (including child abuse and murder) and suicidal inclinations to asthma attacks and excessive weight gain—all related to premenstrual syndrome.

When she traced the history of these tendencies, she found coincidentally that in every case they had begun at puberty with each woman's first menstrual period. Many of these women had been on medication deemed appropriate at the time for their particular manifestation of symptoms, but it was not until they switched to supplemental progesterone that relief came almost immediately. Some, even those who had been in prison, no longer had to be institutionalized.[3]

As for menopause, historically far too little has been done to help women through what is for some a severe illness—especially when one considers osteoporosis, a very crippling disease and the result of untreated hormone imbalance. And medical science still wants to set PMS aside as a mystery, rather than a reality to be dealt with. As Dr. Stuart Berger states in his book *What Your Doctor Didn't Learn in Medical School*, PMS "has managed to remain something of a medical enigma."[4]

The PMS and menopause conundrum does, however, continue to be discussed and written about in medical publications. There appears to be a consensus that perhaps 40 to 60 percent of women under the age of fifty are affected to some extent by PMS.[5] For women in the childbearing age range, it is somewhat less common, with perhaps 20 to 40 percent afflicted. But fully a fourth of these, 5 to 10 percent, have such severe PMS that it significantly disrupts their lives.[6]

It is interesting to note Dr. Katharina Dalton's comment that "target cells containing progesterone receptors are widespread in the body, although most are found in the brain, particularly in the limbic area [near the brain stem], which is the area of emotion, rage and violence."[7, 8] The body's other receptor sites for progesterone are in the eyes, nose, throat, lungs, breasts, liver, adrenals, uterus, and vagina.[9, 10] "All these," says Dr. Dalton, "are areas in which symptoms of PMS may occur such as headaches, asthma, laryngitis, pharyngitis, rhinitis, sinusitis . . . mastitis, alcohol intolerance and congestive dysmenorrhoea." In fact, as many as 150 symptoms throughout the body have been recorded that relate to PMS.[11]

Carol Petersen, R.Ph., of Women's International Pharmacy, says that around menopause or when symptoms escalate, the estrogen dominance that is at the root of all this is often intensified by the introduction of synthetic progestins such as Provera, for they block the brain receptors from receiving natural progesterone. Dr. John R. Lee describes how such an imbalance can wreak havoc even without drug intervention, simply as a result of a woman's deficiency of natural, balancing progesterone:

> Low premenopausal progesterone, as a consequence of anovulatory cycles, can induce increased estrogen levels and lead to symptomatically significant estrogen dominance prior to menopause. The most common age for breast or uterine cancer is five years before menopause. And there is more. The hypothalamic biofeedback mechanism activated by this lack of progesterone as a woman approaches menopause, leads to elevation of GnRH [gonadotropin-releasing hormone] and pituitary release of FSH and LH. Potential consequences of this are increased estrogen production, loss of corticosteroid production, and intracellular edema. Heightened activity of the hypothalamus, a component of the limbic brain, can induce hyperactivity of adjacent

limbic nuclei leading to mood swings, fatigue, feelings of being cold, and inappropriate responses to other stressors . . . . Hypo-thyroidism is suspected despite normal thyroid hormone levels.[12]

What Causes Severe Premenstrual Cramps?

Speaking from my personal experience—life could have been so sweet if I'd had a glance into the future! But then again, it might not have been as enlightening or purposeful. The struggle to reach our goals is often as rewarding as their final attainment. Even uncovering small pieces of information that eventually lead up to the big picture can be fulfilling.

One such piece came to me in the mail in a medical report from which I learned that cramping starts when the adrenal gland has been drained of its cortisone reserve. Another article came from the *Cancer Forum*, published by the Foundation for Advancement in Cancer Therapy, presenting Dr. Lee's findings. It helped me understand that progesterone is a precursor of cortisone, which is made by the adrenal glands.[13]

In reading about this, I could easily remember suffering from severe cramps as a teenager. The pain would be so intense that wherever I was— at work or school—I would often faint. I would end up in the clinic for the rest of the day with a hot water bottle, hot tea, and aspirin every two hours. As women of menstrual age continue to have these problems, it's appalling that education on natural solutions is not available from most of our doctors. Instead, unnecessary suffering continues, and medical doctors continue to prescribe the common synthetic drugs.

Yet, many women have found progesterone to be a pain-relieving hormone. Cramping at the onset of a period can be painful and disruptive, but progesterone helps alleviate the discomfort by assisting the adrenal glands to create cortisone. According to Betty Kamen, Ph.D., some physicians now advise the "application of one-half teaspoon of the cream to the abdomen every 30 minutes until cramping subsides."[14]

Anecdotal testimony comes from Dr. Linda Force, who tells me that prior to using the progesterone cream, she had a problem with clotting during her periods. But administering the cream has given her a normal, even flow and kept her periods regular. She now applies it every morning and evening, right up to the time her period begins. When her period stops, she starts all over again.

Be aware, however, that many doctors fail to associate our symptoms with PMS or menopause. They do not typically recognize and acknowledge our difficulties as being related to progesterone deficiency, but rather attempt to treat only our symptoms.[15] An example of this can be seen with many postmenopausal women who are not aware of the significance of the decline in their progesterone levels. They are finding that even though they eat low-fat foods, their cholesterol levels have become elevated. Since they are often put on *unopposed* conjugated synthetic estrogen, their LDL (less desirable) cholesterol has consistently risen. This is the result of estrogen dominance. Instead of neutralizing this dangerous hormone with natural progesterone, the doctor will often prescribe one of the many drugs that lower cholesterol. Meanwhile, the estrogen in their bodies remains unopposed and continues to be threatening.

It is vital to keep our thoughts focused on the natural alternatives to these drugs. In appendix F and appendix G of this book you will learn how to have natural progesterone prescribed by your doctor, or where you can obtain natural progesterone in a variety of forms without a prescription.

It's comforting to realize that experts have found a combination of all-natural ingredients that will work to create a proper balance between a woman's own estrogen and progesterone. Replacement of the body's own natural progesterone addresses any deficiency, and the result is relief from many worrisome symptoms.

PREMENSTRUAL EPILEPSY AND DEPRESSION

Dr. Ray Peat has detailed studies showing that when epilepsy occurs prior to menstruation, it is often relieved by progesterone therapy. This therapy has also been used with success in suicidal depression, Reynaud's phenomenon, Meniere's disease (inner ear), kidney disorders, and abnormal liver metabolism.[16]

This has been validated by Dr. Dalton, who says, "One of the most satisfying experiences is to diagnose and treat a woman with premenstrual epilepsy. She can be treated with progesterone and freed from all anticonvulsant tablets with their many and unpleasant side effects."[17] Dr. Betty Kamen, in her book *Hormone Replacement Therapy: Yes or No?*, concurs that progesterone has an effect on epileptic seizures because of its barbiturate-like action on brain metabolites.[18]

This item, which appeared on the Internet, seems to corroborate those statements. It came in from a woman who had suffered in the past with epilepsy. She wrote: "Many years ago, at my absolute worst, I was having 30–50 seizures a day. Since I went on 200 mg. of natural progesterone (capsule form) a day, I have been nearly seizure-free. I know the vitamins and nutrients that I'm on are also helping."

Dr. Dalton's experience and study make clear that many of the uncomfortable symptoms normally associated with a woman's monthly cycle occur just prior to and during the first few days of menstruation, and occasionally at ovulation. It is not uncommon to experience pain, depression, and headaches continuing through the first day or two of each menses. However, all these ill effects, often aggravated by stress and its consequential depletion of progesterone, can be bypassed. Once the progesterone is replaced in our bodies naturally, many of our problems clear up. We *can* avoid the sufferings of hormone deficiency, and whether in our teens or in the postmenopausal period, we can be thankful for the efforts of Dr. Katharina Dalton.

PROGESTERONE, STRESS, AND THE ADRENAL GLANDS

Even though circulating progesterone is produced in the ovaries, Dr. John Lee says that progesterone is also manufactured by the adrenals (our "stress glands"), where it is converted into the corticosteroid hormones. This progesterone is immediately and continuously used to supply the multitude of adrenal functions.

Niels H. Lauersen, M.D., says that "when natural progesterone drops, the normal conversion by the adrenal glands cannot take place, salt may build up, fluid may be retained, and hypoglycemia may ensue. Synthetic progestins generally make PMS symptoms worse, so if a woman is about to be treated with progesterone, she should be sure that it is natural progesterone."[19] According to Dr. Lee, this will balance out the estrogen dominance that is contributing to these symptoms.

Dr. Robert Lindsay, an osteoporosis expert, confirms that the synthetic hormones cause an increase in stress. He says, "If [they're] given ten milligrams of Provera in combination with ... Premarin, many women feel premenstrual and crabby and irritable. They call up and say, 'Why did you give me that stuff?'"[20] Natural progesterone, however, not only protects against hypertension, as reported in 1990 by the *Journal of Epidemiology*, but also has an antistress effect on the pituitary.[21] This does not appear to be true with synthetic hormones, which increase cellular sodium and may lead to increased hypertension.[22]

Sterling Morgan writes that progestins can "cause temporary hypoglycemia by blocking adrenal production of glucocorticoids, which regulate blood sugar."[23] It's understandable that women so often decide to discontinue synthetic hormones. "Such problems," the author says, "are practically unknown with the natural kind [of progesterone] since it is totally compatible with the human body."[24] And Dr. Peat explains how progesterone aids in many allergic diseases, including the autoimmune and "collagen" diseases. It does so by helping to maintain blood sugar levels and by stabilizing lysosomes—cellular components that are involved in the inflammation process.[25]

Dr. Lee says that women in their thirties (some even earlier, and long before actual menopause) will on occasion not ovulate during their menstrual cycles. From my understanding, this can occur after a vigorous athletic training schedule, trauma, injury, harsh dieting, use of some methods of hormonal contraception, or severe emotional stress. Dr. Lee elaborates:

> Without ovulation, no corpus luteum results and no progesterone is made. Several problems can result from this. One is the month-long presence of unopposed estrogen with all its attendant side effects leading to the syndrome known as PMS. Another is the present, generally unrecognized, problem of progesterone's role in osteoporosis. Contemporary medicine is still unaware that proges-terone stimulates osteoblast-mediated new bone formation. . . . A third is the inter-relationship between progesterone loss and stress. Stress influences limbic brain function including the functioning of the hypothalamus.[26]

Dr. Lee makes clear that *stress can cause missed ovulation.* A decline in progesterone disrupts the production of adrenal corticosteroids. Stress is thereby heightened, causing a woman to be vulnerable to anovulatory cycles.[27]

We often hear about the physical stress caused by excessive exercise. Studies show that stress endured by long-distance marathon runners causes a loss of 4.2 percent of their bone mass in one year. This is due to stress factors that inhibit ovulation, and thus progesterone production.[28, 29] Dr. Peat explains that toxins build up at this time and can also contribute to a progesterone deficiency. Needless to say, the entire body can be affected by this hormone imbalance.[30]

Raising DHEA Levels with Phytohormones

Stress can also deplete another of our vital hormones produced by the adrenals: dehydroepiandrosterone (DHEA). This is the most abundant hormone in the human body, and yet until recently it has received the least attention from medical science! Norman C. Shealy, M.D., a neurosurgeon, has been doing research and giving seminars on the importance of DHEA in men and women. He says that the normal range in men is 180 to 1250 ng/dl (nanograms per deciliter) and in women 130 to 980 ng/dl. Dr. Shealey has found that the patients who fall in the low normal range and below tend to have adrenal exhaustion. These individuals cope poorly with stress, and many have physical and mental disabilities (often serious diseases).[31] According to Dr. Alan Gaby, as well as others, low levels of DHEA have been associated with premature aging, breast cancer, osteoporosis, rheumatoid arthritis, heart disease, obesity, diabetes, poor immunity, and even lupus and Alzheimer's disease.[32, 33]

Preliminary studies suggest that DHEA and progesterone have a heretofore unsuspected link, and that the use of natural progesterone cream may safely help DHEA levels to stabilize. Dr. Shealey and other researchers believe that in most cases this approach is far better than taking synthetic DHEA supplements, as it allows the body's wisdom to determine just how much of the hormone to make.[34] In extreme cases, sometimes DHEA is given for temporary relief, along with progesterone for balance. But a young person taking DHEA simply for increasing energy might damage her adrenals by signaling them not to produce any more of the hormone.

A CONVINCING ARGUMENT

Parallel research indicates that *Dioscorea,* the wild yam plant, contains natural precursors to DHEA. The *Journal of Clinical Endocrinology and Metabolism* states that higher levels of DHEA have been associated with a more active immune system and a decrease in diseases of the major organ systems.[35] Joe Glickman, Jr., M.D., refers to various studies showing that people with low DHEA levels are more prone to hardening of the arteries, strokes, and cancer. In a twelve-year study of more than 240 men, those who had higher DHEA levels had a lower death rate from all these diseases.[36–40] Dr. Glickman even predicts that DHEA "may prove of significant benefit in treating AIDS."[41]

Raising DHEA levels may also reduce gastric cancer, as well as have a general anticarcinogenic effect on the breast, lung, colon, thyroid, skin, and liver. [42–46] Moreover, the *Journal of Neuroscience Research* reports that DHEA "greatly increases the number of nerve cells and their ability to establish connections to other nerve cells," suggesting that DHEA levels may help prevent Alzheimer's and other age related senility diseases.[47]

The important factor to remember here, says Dr. Glickman, is that in its synthetic (prescription) form, DHEA could have damaging side effects to the liver.[48] Using a natural plant source seems to be the key to the many benefits of this hormone—especially in prevention of diseases brought on largely by the wear and tear of chronic stress to the system. Dr. Neecie Moore informs us that we can beneficially raise our DHEA levels (just as we can our progesterone levels) with the precursor found in the wild yam. She illustrates with the story of one doctor who raised his DHEA levels 91 percent in sixty days with *Dioscorea*.[49] (See chapter 5.)

The more I have learned about natural progesterone, the more intrigued I have become with it. In case you are debating whether or not to take natural progesterone and wondering whether it is really necessary, Dr. Ray Peat gives some more reasons: "When progesterone is deficient, there tends to be hypoglycemia," he says. Furthermore, natural progesterone has a stabilizing action in the muscle tissue and many other areas of the body, "such as the uterus, blood vessel walls, the heart, the intestines and the bladder. Less visibly, progesterone stabilizes and normalizes nervous, secretory and growth processes. Biochemically, it plays a major role in providing the material out of which all the other steroid hormones (such as cortisone, testosterone, estrogen and salt-regulating aldosterone) can be made as needed."[50] In short, he says, progesterone plays a multitude of roles in normalizing body functions.

The Right and Wrong Ways to Ease the Transition

Now let's explore further some of the evidence as it relates to menopause—keeping in mind that most doctors deal with this critical period in a woman's life with a well-known prescription drug, the use of which has recently become more and more controversial. The synthetic estrogen known as Premarin is given to more than twenty-two million women per year and is one of the ten most prescribed drugs.[51] And the prescribing of Provera (medroxyprogesterone), touted as being for "prevention," is keeping pace. How sad, when the truth of the matter is that both are synthetic hormones that are very difficult for the body to metabolize and thus are capable of causing extreme reactions.

Women Who Speak Out

In a personal account, Gail Sheehy elaborates on this point. She tells us in her book *The Silent Passage: Menopause* what happened when she added synthetic Provera to her regimen:

I felt by afternoon as if I had a terrible hangover. This chemically induced state was not being subdued by aspirin or a walk in the park. It only worsened as the day wore on, bringing with it a racing heart, irritability, waves of sadness, and difficulty concentrating. And to top it off, the hot flashes came back! . . . . By night I couldn't go to sleep without a glass of wine, and even then was awakened by a racing heartbeat and sweating."[52]

It's frightening to learn that most of our traditional medical doctors prescribe progesterone's counterfeit, Provera, despite the fact that, Sheehy points out, "it has never been approved for treatment of menopause by the U.S. Food and Drug Administration."[53] She continues:

Notwithstanding, the FDA's Advisory Committee on Fertility and Maternal Health Drugs stated in '91 that this combination of hormones "may be used indefinitely by a woman with a uterus." Asked what proportion of the female population over age fifty would be suitable candidates for long term consumption of estrogen alone or combined with synthetic progesterone, the committee replied, "virtually all." A blank check.[54]

Some of these synthetic hormones have been on the market for approximately fifty years. However, year after year since then, women have come home with their prescriptions and continued to have the same miserable reactions and apprehensions about taking them. Many never even fill their prescriptions, and some who do will not have them refilled, especially after reading the manufacturers' warnings (see appendix B). We are afraid and often don't know where to turn. Gail Sheehy expresses what many women have experienced:

I didn't require a ten-year clinical trial and double-blind study to guess what was going on. Taking synthetic progesterone with the estrogen for half of each month was like pushing down the gas pedal and putting on the brakes at the same time, and it had left my body confused and worn out.[55]

Before we go into some of the conditions that accompany menopause, keep in mind the numerous physiological benefits of natural progesterone—versus the associated problems that are known to be aggravated by use of synthetic progestins, such as cardiac insufficiency, epilepsy, migraine headaches, depression, high cholesterol levels, and kidney disorders.[56] When you ask your doctor for natural progesterone in lieu of the synthetic substance, he may say that the artificial hormones have been proved to mimic our natural hormones perfectly. If so, refer your doctor to an article in the *New England Journal of Medicine* that appeared in 1993.[57] Remember that this information, urgently needed by women, was discovered by scientists who held themselves above the politics of medicine. Dr. Betty Kamen's words aptly summarize this news:

> Progesterone is the primary building block for all the other steroid hormones. This alone distinguishes natural progesterone from synthetic progestins. . . . The one similarity between the synthetic and natural forms of progesterone is that . . . each can trigger uterine bleeding similar to menstrual flow, if needed.[58]

As we highlight the many benefits that develop with natural progesterone supplementation, I must relate to you a surprising turn of events described to me by a friend. I had told her Dr. Linda Force's story of how the application of progesterone had cleared up her fibrocystic breast disorder. My friend then decided to use the natural progesterone cream on the lump on her breast, which at one time she'd had to have drained. This fibroid of hers would come and go, and at this point she was not overly concerned because she thought that with proper diet and exercise it would eventually go away.

However, after thinking about progesterone therapy for a while and learning that it was natural and had no side effects, she decided to use the cream as a preventive measure and apply it directly to the lumpy area on her breast. When we met the following month, she was elated with the results of the progesterone cream. She told me that after the first few days of using it, she began to experience other unexpected benefits. Not only did the lump go away but *so did her hot flashes*. This was a surprise to her, because everything she had read indicated that supplemental estrogen was the only remedy for hot flashes.

Of course, if she'd had the opportunity to read any of Dr. John Lee's writings, my friend would have understood that a high percentage of women with hot flashes respond to the progesterone alone. The additional need for estrogen (preferably estriol) is indicated only if the hot flashes do not subside within about six months of the beginning of progesterone treatment, and it should then be a very individual and informed decision. Every body is different and will experience benefits in its own unique way. My friend also mentioned that her insomnia was no longer a problem. As long as she used the progesterone cream, she was able to sleep through the night without the usual interruptions that are brought on by hormone imbalance.

Another friend, Dianna Widerstrom, shares her experience: "Two weeks before my period, my breasts would swell so badly that they would be painful to the touch or whenever I lay on my stomach. (I had been diagnosed as having fibrocystic breasts six years ago.) When I finally started my period, I would pass blood clots the size of a silver dollar. At this time I also experienced a tremendous amount of water retention. During all this stress to my body I would also become very irritable.

"As I was relating my problems to a dear friend, she told me about progesterone cream. I began using the cream, and after the first month I saw several changes in my body—the breast lumps had reduced tremendously; the clotting wasn't as bad, nor was the water retention. After three months, the lumps were completely gone and the blood flow had become even. All these things have had a positive effect on my personality, which hasn't gone unnoticed or unappreciated by my husband.

"After several more months I decided to take a month off the cream. What a mistake! I'll never do that again! My breasts became lumpy and very sore. I was passing blood clots again (though not as large as before); my periods no longer had a nice even flow, and I retained a lot of water. This all proved to me that my body needs and works well with progesterone. (I've also discovered that dandelion root extract works well to reduce water retention.)

"I am a salon owner and have decided to sell the cream in my studio. Several of my clients use it and have seen major changes in their bodies, and thus their lives. Unfortunately, we are not always able to obtain these natural ingredients from our doctors."

Hot Flashes, Breast Swelling, Vaginal Atrophy, and Infections

Estrogen is produced by the ovaries and also stored in body fat. When it is needed, it is released and then metabolized by the liver.[59] As we get older, the body tells us when our hormones are declining. According to the experts, the two major signs of low estrogen are hot flashes and genital atrophy. Although Dr. Lita Lee says, "I have never seen a menopausal woman (whether natural or surgical) who did not successfully ameliorate hot flashes with progesterone treatment,"[60] I have personally talked with several menopausal women who experienced an *increase* in breast swelling, water retention, insomnia, or headaches during the *initial* stages of progesterone treatment.

I asked Dr. John Lee why this might be. He explained that before these women started their progesterone therapy, they were estrogen-dominant. Estrogen dominance tunes down the estrogen receptors. The body is trying to protect itself from too much estrogen. Progesterone, however, temporarily increases the sensitivity of the receptors. Thus, for a short duration one might experience estrogen-like side effects, such as breast swelling and bloating, which in time (after a two- or three-month period of progesterone application) usually subside.

He illustrated that a similar thing happens with a woman who is in her premenopausal phase. She is still making, as he put it, "a ton of estrogen" and is menstruating regularly; but she has no progesterone to bring about orderly periods, so they are sometimes heavy and sometimes light. The body is now making more estrogen than it really needs, because the receptors are not very attentive to it. When you then add progesterone, you restore estrogen dominance for a while as the receptors resume normal sensitivity. At this time there may be a period of adjustment when some women experience various uncomfortable symptoms; all that estrogen they are making may suddenly flare up as swollen breasts, weight gain, water retention, and headaches.

These estrogen side effects last only for the first month or two, if they're even noticed at all. By the third month, the body's progesterone level is sufficiently high to oppose this action. Dr. Lee's suggestion to women with these problems, therefore, is that if you will just persevere for two to three months, your progesterone level will rise enough to do its beneficial work and the temporary side effects should disappear.

Another common problem associated with the menopause is vaginal atrophy, which can be extremely painful with intercourse. Dr. Carlton Fredericks suggests that the use of vitamin A and E cream for atrophic vaginitis is much safer than the synthetic estrogen creams (Cynonal or Premarin) so frequently prescribed by medical doctors. He says, "Both have side effects and may also irritate you."[61]

If your doctor thinks your symptoms indicate a need for estrogen treatment, remember that natural progesterone should be the first treatment for at least two or three months. Applied topically, it is very effective for many women. Others whose problem has persisted have achieved results with estriol cream. When the ingredients essential to the health of the vaginal tissue begin to decline, the genital tissue often degenerates and becomes vulnerable to infection.

Dr. Lee also talks about this potential for vaginal dryness and "atrophic mucosa" (shrinking membranes) after menopause. The condition will in turn "predispose women to vaginal, urethral and urinary bladder infections.

. . . To treat the infectious agent," he says, " . . . with antibiotics is only temporarily successful because the underlying real cause . . . is loss of *host resistance* secondary to hormone deficiency. . . . A recent controlled trial of intravaginal estriol in postmenopausal women with recurrent urinary tract infections found that estriol significantly reduced the incidence of urinary infections. . . ."[62]

Dr. Lee says that in his clinical practice, estriol treatments have "resulted in the re-emergence of friendly lactobacilli [bacteria] and the near elimination of [undesirable] colon bacteria, as well as the restoration of normal vaginal mucosa and a resumption of normal low pH (which inhibits the growth of many pathogens)."[63] Where estrogen had been contraindicated to his patients, Dr. Lee was surprised to find that the natural progesterone therapy also cleared up the vaginal dryness and mucosal atrophy after three to four months of use. His experience and that of many other doctors is confirmation of our need to find the right individual balance of estrogen and progesterone in their *natural* form.

How Do Drugs Affect Our Health and Hormone Levels?

Dependency on medication is a disease in itself, as it leads to an addiction to many drugs. The average American may think at first, "I'm not dependent on medicine!" However, reality strikes when we take a look inside the medicine cabinet. For example, consider those antibiotics prescribed so frequently and without consideration of their long-term effect. Because antibiotic therapy results in the destruction of useful bacteria, it creates an ideal environment for the overgrowth of more harmful bacteria or fungi (yeasts).[64] Consequently, other infections take over and virulent, drug-resistant strains of many disease-causing organisms evolve. Dr. Julian Whitaker says, "13,300 hospital patients in 1992 died from infections that could not be controlled by antibiotics."[65]

We might well turn our attention from *anti*biotics to *pro*biotics—a word used in the informative book *Alternative Medicine*. Providing our own friendly bacteria with proper nutrition will eventually create and assure an excellent state of homeostasis. According to this book, some of the probiotics include *Lactobacillus acidophilus, Lactobacillus bulgaricus, Bifidobacterium bifidum,* and *Bifidobacterium longum.*[66] While antibiotics such as penicillin kill our good bacteria, and steroids such as cortisone and birth control pills may also harm the flora of the bowel, these beneficial bacteria can thrive on live yogurt and diets that are rich in complex carbohydrates (organic raw vegetables, whole grains, legumes) and low in sugar and fats.[67]

Many women suffer from an overgrowth of *Candida albicans*, a fungus common to all of us and harmless when kept in check by friendly bacteria. Often difficult to control, this condition is more widespread than we generally recognize. Its symptoms may include chronic vaginal irritation, thrush, an itchy skin rash, headaches, extreme fatigue, and problems of the gastrointestinal or urinary tract or the neuromuscular or respiratory system.[68, 69]

Frequently brought on by antibiotic therapy, candidiasis may also be related to the use of contraceptive and other corticosteroid drugs as well as to dietary factors (e.g., excessive sugar) and to progesterone deficiency. The amount of sugar in many women's diets not only feeds the yeast but upsets our hormone and mineral balance. Dr. Lee brings the point home:

> One can of coke contains nine teaspoons of sugar. Dumping nine teaspoons of sugar into your body is a setup for wildly fluctuating blood sugar levels, weight gain, insulin resistance, and adrenal fatigue—a perfect setup for hormone imbalance? Just as bad are the diet sodas with aspartame [NutraSweet], a synthetic chemical containing substances called excitotoxins, which are known to cause brain damage and may contribute to hyperactivity, learning disabilities, and Alzheimer's.[70]

To treat yeast infection and the associated uncomfortable discharge, doctors often prescribe one of several fungicidal creams, such as Monistat-7 or similar over-the-counter drugs. Thinking of my own experience with these products stirs up an unpleasant memory of vaginal tissue burning and irritation. But having this additional side effect to compound my other menopausal troubles would not have been necessary, had I known then about *natural* hormone treatment.[71]

Natural Hormones Raise "Good" Cholesterol and Lower the "Bad"

Another area of concern relates to our cholesterol levels as we age. Have you had yours checked lately? If you are menopausal, have you noticed that it's been higher than it used to be and that all those "cholesterol-free" products don't seem to be having any effect on lowering the "bad" cholesterol?

Diet and exercise are certainly important, but they are not the whole truth in controlling cholesterol. Gail Sheehy also suggests a relationship between hormones, cholesterol, and cardiac disease:

> The most significant predictor of heart disease is the HDL level. Bad cholesterol levels normally increase in women for some ten to fifteen

years following the cessation of periods. Again, dangerous changes in cholesterol count or blood pressure do not announce themselves with obvious symptoms, not until there is a medical catastrophe. "If your HDL level is low, and your LDL level is relatively higher—even if you're walking around with a total cholesterol count of 100—you're going to be in trouble," says Dr. Estelle Ramey.[72]

As you may recall from our discussion in chapter 2, clinical experience emphasizes the need for stabilizing estrogen with progesterone to counter the risks of lowering our supply of HDL cholesterol. Conventional estrogen treatment, by itself, can cause many harmful side effects and further accelerate this problem.

Luteal Phase Defect

A common complaint of many women today is, "My periods have completely stopped" or, "My periods are abnormally frequent or irregular." Medically, these symptoms are referred to as "luteal phase defect." Explained simply, estrogen is the dominant hormone during the first part of the cycle (the follicular phase). But if it continues to predominate in the second part of the cycle (the luteal phase), this indicates low progesterone production and could bring about any of these menstrual abnormalities, referred to as hypermenorrhea, polymenorrhea, or amenorrhea.[73]

Hypothyroidism

As Dr. Lee has said, low progesterone is often misdiagnosed as thyroid deficiency. Nevertheless, Dr. Peat emphasizes that thyroid hormone is basic to all biological functions and that sometimes both thyroid and progesterone supplements are needed, as "each has a promoting action on the other." To see whether thyroid supplementation might be needed in addition to the progesterone, he recommends a test called the Achilles tendon reflex, which measures muscle energy by the speed at which the calf muscle relaxes.[74]

"Without adequate thyroid," says Dr. Peat, "we become sluggish, clumsy, cold, anemic, and subject to infections, heart disease, headaches, cancer, and many other diseases and seem to be prematurely aged. . . . Foods aren't assimilated well, so even on a seemingly adequate diet there is 'internal malnutrition'."[75] Irregular periods, often leading to needless hysterectomies, are common aspects of hypothyroidism; and breast disease, he says, is another classic manifestation. In explaining this, Mark Perloe, M.D., says, "Too little thyroid production may cause . . . increased prolactin levels and persistent estrogen stimulation."[76]

In a conversation, Dr. Peat told us that estrogen (which we can try to balance with supplemental progesterone) inhibits the release of thyroid hormone from the gland, whereas an adequate amount of thyroid hormone raises natural progesterone production and lowers estrogen. That makes it easy to see how thyroid hormone and progesterone can complement each other. He even made the interesting observation that since estrogen and cortisone weaken the blood vessels, progesterone (along with thyroid supplements) is a good way to help prevent easy bruising.

Unfortunately, our physicians often fail to understand or explain the benefits of natural (marketed under the name "Armour") over synthetic thyroid medication. Though the formula has changed somewhat in recent years, Dr. Peat calls the natural "the most generally effective," since "many people whose thyroid gland is suppressed by stress cannot respond to synthetic thyroxine, T_4." Additionally, those whose liver function is not up to par may require the active form of the hormone, T_3, to which thyroxine is normally converted in the liver. While an excess of unconverted T_4 can be toxic, sometimes just a bit of carbohydrate between meals (to maintain a stable level of glucose in the liver) allows the conversion to T_3.[77]

Contrary to the advice given in most "health books," Dr. Peat counsels against an excess of dietary iodine, which might depress thyroid function. He has also done some groundbreaking research that indicates that an excess of *unsaturated* oils in our diet powerfully inhibits the thyroid. To counteract the resulting fatigue, obesity, and other effects, he advocates the regular consumption of coconut oil (see chapter 6).[78]

Dr. Peat has also concluded, on the basis of convincing research, that hypothyroidism (along with too much estrogen) is the main cause of multiple sclerosis (MS), and that progesterone deficiency is one of several other factors that may be involved. The mechanism appears to be that edema caused by the unbridled estrogen can result in blood clots in the brain, with associated areas of destruction of the myelin sheath of the nerve tissue.[79] Low levels of DHEA have also been linked to MS.[80]

Who Needs Premature Wrinkles or Other Skin Problems?

Dr. Peat has somewhat encouraging news for those of us concerned about wrinkled skin. In spite of estrogen's promotion as a "youth drug," he says it "is known to advance the aging of collagen in all tissues that have been studied, including skin." He comments wryly that "women, like cows, will puff up with water and fat under the influence of estrogen, and wrinkles will naturally be smoothed out, but the skin itself is actually losing its elasticity faster when estrogen is used." On the other hand, he says, "Progesterone has been found to reverse the chemical changes which occur in collagen with aging"—as well as to normalize the immune system so as to suppress reactions that may contribute to the aging process.[81]

But one should not expect miracles! This treatment seems to be more effective for some women than for others and is probably dependent on many factors, including the age at which it is begun and the degree of prior overexposure to the sun. Dr. Peat, however, has one more bit of advice: use coconut oil in place of skin creams and lotions containing the much-touted polyunsaturated botanical oils. These latter, he says, actually promote aging of the skin by intensifying the effects of the sun's ultraviolet rays!

We know of other skin conditions as well that sometimes respond to topical progesterone therapy. Dr. Lee can document cases where his patients' acne, psoriasis (scaly reddish patches), rosacea (rose-colored, flaky areas near the nose and forehead), seborrhea (flakes and itching), and even keratoses (hardened skin cells that could be precursors of squamous-cell skin cancer) cleared up when they applied progesterone cream. This natural approach sounds wonderful compared with other options, such as surgery or antibiotics (which kill friendly bacteria, often leading to "leaky gut" syndrome and candida yeast overgrowth).[82]

WHY DO I HAVE THESE MUSCLE ACHES AND PAINS?

In evaluating symptoms of illnesses common to women, I found it interesting to note how many diseases elicit complaints similar to those of PMS and menopause. For instance, a popular diagnosis given to an incredible number of female patients is *fibromyalgia*. As I was interested in the subject, I decided to attend a lecture given by a well-known rheumatologist from Emory University. In his talk he mentioned that fibromyalgia is a disease found predominantly in women. And sure enough, the huge assembly room was packed with women of all ages.

A couple of months prior to this event, I had met a woman at a senior citizens' health fair who told me she had fibromyalgia. Because I was researching a book on arthritis, I had learned that a physician from the Mayo Clinic had named this disease "fibrositis" and called it "the most common form of acute and chronic rheumatism." The report continued, "Its symptoms are similar to chemical sensitivities and chronic fatigue syndrome."[83] It turns out that other doctors and medical studies report that "a number of poorly understood conditions, such as chronic fatigue syndrome or fibromyalgia, are misdiagnosed as 'chronic Lyme disease.' Fibromyalgia occurs mostly during midlife but it may be seen at any age."[84] In reviewing an article in the *Journal of Internal Medicine*, Betty Kamen says, "There is no evidence that fibromyalgia is a disease of the muscle or a rheumatic syndrome."[85]

What caught my attention most is that the symptoms mimic all the characteristics of progesterone deficiency or estrogen dominance that we list in chapter 2: depression, fluid retention, joint and muscle pain, headaches, insomnia, poor memory, and even PMS.

So my newfound friend decided to use the progesterone cream. After approximately one month, she called me and excitedly announced that all her "fibromyalgia" symptoms were gone. She was particularly happy when her distended abdomen began slimming down and she had to buy new clothes a couple of sizes smaller. This woman was not afraid to speak up when it was necessary. So I wondered what her doctor was going to say when she told him, "Well, doctor, after using this botanical progesterone cream my symptoms just disappeared." In my mind I could just hear her doctor saying, "Ahaa—ahaa...mmhmmmm—mmhmmm. . . . Sure, lady— and I know a good psychiatrist for you, too."

But this woman knew she was onto something. For over thirty years she had been living with these problems, and the only times she had been completely symptom free in her married life was when she'd been pregnant (seven times). As Dr. John Lee tells us, "In pregnancy, the placenta produces 300–400 mg of progesterone a day in the third trimester." (This is when many women feel the healthiest.) This amount is approximately ten times higher than what's normally produced in an ovulating woman.

What great protection progesterone provides us during pregnancy! It's the hormone essential for the survival of the embryo. *Could it also be essential for the survival of our good health?*

IF I'D ONLY KNOWN THEN WHAT I KNOW NOW (POSTMENOPAUSE)

Once a woman has reached the stage in her life when she no longer has natural menstrual periods, why must she be artificially thrust back into her cycle? How unrealistic that women should endure these drug-forced cycles and suffer through multiple side effects in lieu of passing through what should be a peaceful time of life. I have found that natural hormone replacement therapy will not bring on a period when taken on a regular basis. If I had known back then, at the peak of menopause, what I know now, I would have put the knowledge into practice immediately.

In her book titled *The Menopause Industry: How the Medical Establishment Exploits Women*, author Sandra Coney talks about how the pharmaceutical industry has promoted the idea of menopause as an ailment because it has the drugs to medicate the symptoms. By treating menopause like a disease, we condition the body to sustain monthly periods synthetically and indefinitely—right into old age.[86]

What has been promoted by the pharmaceutical companies as a dream to preserve youth has turned out to be more of a nightmare, however. Promises of synthetically covering up natural menopause and postmenopause with estrogen have far from lived up to expectations. And the reality of this nightmare is showing up in the studies. The *American Journal of Obstetrics and Gynecology* now states that not only long-term, but short-term users of HRT carry a 40 percent risk of acquiring breast cancer.[87]

Through preventive care, it is possible that the pre-, peri-, and postmenopausal years can turn out to be very constructive ones. We can devote our energies to achieving a degree of health that allows us to reach out to others with stamina and creativity. As Margaret Mead, author and anthropologist, puts it, "The most creative force in the world is the menopausal woman with zest."[88]

New Thoughts About Contraception

But what about those still in their reproductive years? Statistics show that ten million women in the United States[89] and as many as fifty million worldwide are using oral contraceptive agents.[90] And the average layperson is confused as to whether it is progesterone or estrogen that is prescribed for contraception. According to *Goodman and Gilman's The Pharmacological Basis of Therapeutics*, "Progestational agents . . . in combination with estrogens are used widely as oral contraceptives." Ethinylestradiol, mestranol, and many other combinations are listed.[91] However, Dr. John Lee tells us that the commonly prescribed synthetic contraceptive ethinylestradiol is considered dangerous "because of high oral absorption and slow metabolism and excretion [which are] true of all synthetic estrogens."[92]

The long-term side effects are often tragic. Since progestational agents were introduced in the 1950s, the number of synthetic forms of the drugs has increased abundantly, and many are used clinically as contraceptive agents. Norethynodrel was one of the very first progestin compounds to be used. Among other highly potent progestins are chlormadinone acetate and cyproterone acetate.[93] The U.S. Food and Drug Administration issued a warning concerning medroxyprogesterone, one of the most commonly prescribed and one that seems to cause the most agitation and discomfort. Evidently the agency had not approved it for contraceptive use because of "potential serious side effects."[94] Nevertheless, it continues to be used.[95]

Scientists have studied the action of progesterone receptors in animals. They've demonstrated, as Dr. Katharina Dalton attests, that "progesterone receptors will only transfer molecules of progesterone to the cell nucleus, but not molecules of the artificial progestogens. . . . [Yet] these substitutes for progesterone are used in the contraceptive pills and in other gynecological conditions."[96]

Dr. Ray Peat cautions that the use of estrogens, birth control pills, and even I.U.D.s can cause progesterone deficiency.[97] This in turn can cause severe cramping, infertility, weight gain, and other seemingly unrelated problems. In fact, breast cancer is included among the long-term reactions to birth control pills.[98] He also directs our attention to a 1966 study by Spellacy and Carlson that demonstrated the danger of prolonged pancreatic stimulation by oral contraceptives. This increased the level of free fatty acids and the tendency toward diabetes.[99]

The unexpected can occur when the body is exposed to the dangers of chemically altered hormones. As Dr. John Lee tells us, "Some of these drugs [have] resulted in permanent loss of ovary function. . . . [And] vaginitis occurs more often among women taking contraceptive pills. . . . Contraceptive pills prevent the normal hormone-generated mucus from being produced to protect them. Birth control pills work, after all, by suppressing normal hormones."[100]

Oral contraceptives have also been clearly linked with the development of blood clots. Clotting may occur in the veins of the legs—a condition known as deep venous thrombosis. Since blood clots in the leg (calf, thigh, etc.) can easily travel to the lungs (pulmonary embolism), these are quite serious. *The Complete Book of Menopause* states that "pulmonary embolism can be fatal and [is] the basis of the major concern about deep venous thrombosis."[101]

We see time and again the importance of introducing natural progesterone to oppose the dangerous effects of too much estrogen. In place of the artificial and harmful progestins, Dr. Alan Gaby compares the different results produced by the use of natural progesterone. He says:

Whereas progesterone aids in conception and helps maintain pregnancy, progestogens inhibit ovulation and are therefore used for birth control. . . . Many women who have had severe side effects from birth control pills or from other hormone regimens containing progestogens improve when they receive natural progesterone.[102]

Some medical doctors and many women wonder why we have not been using natural progesterone as a contraceptive, since it has never exhibited any serious or long-term side effects. For while progesterone, given from the time of ovulation, maintains pregnancy—given before, in some situations and at the right strength, it may be able to prevent it. (Apparently it is all in the timing—which can be tricky—taking into account individual cycle variations and responses.) This urgently deserves further study. However, at this stage there is a certain amount of disagreement as to the application of natural progesterone to contraception, and just how much guarantees "safety." *Therefore, the discussion that follows is not to be considered a recommendation,* but merely a starting point for more investigation.

When asked about progesterone having contraceptive action in premenopausal women, Dr. John Lee replies that he uses progesterone in cream form mainly to establish normal physiological levels. There are varying potencies of creams on the market; some are probably so mild as to have no contraceptive effect, while others may have the potential to prevent ovulation in some women.

When a woman would like to have a baby, Dr. Lee advises the patient to wait until day twelve or fourteen to start progesterone. This avoids any inadvertent interference with ovulation. If she is not interested in conceiving and is deficient enough in progesterone to warrant a few extra days of treatment each month, he sometimes moves that up to day eight or ten.

Dr. Lee tells us, "If sufficient [natural] progesterone is provided prior to ovulation, neither ovary produces an egg. This inhibition of ovulation is the original mechanism of action of progestin contraception. . . . Similarly, the high estriol and progesterone levels throughout pregnancy successfully inhibit ovarian activity for nine months. Therefore adding natural progesterone from day 10 to day 26 of the cycle suppresses LH and its luteinizing effects."[103]

The reason for taking higher levels of progesterone before ovulation is that at the time of ovulation, when the follicle converts to the corpus luteum and starts producing progesterone at a mammoth rate, the other ovary senses the excess progesterone and immediately halts its own ovulation process. In normal circumstances, both ovaries are stimulated in the first half of the month by the follicle-stimulating hormone, and both start making an egg. When one of them ovulates and starts producing hormones, that's a signal for the other one to stop. If both ovaries ovulated, we'd have fraternal twins quite often because sperm may enter both fallopian tubes. But fraternal twins occur only once out of every three hundred births—meaning that the rest of the time, one ovary ovulates enough ahead of the other ovary to shut it down. That's the basis of birth control pills. While higher doses do stop ovulation, it is now known that the major contraceptive mechanism of synthetic progestins is that even at low doses they occupy progesterone receptors in the endometrium and prevent implantation of the fertilized ovum (egg).

Lita Lee, Ph.D. is one who also believes that natural progesterone is "nature's contraceptive" and can . . . prevent pregnancy without harmful side effects.[104] Nonetheless, I must share with you a conversation I had with Wallace Simons, R.Ph., President of Women's International Pharmacy, concerning birth control (or to use his terminology, "ovulation control"). Referring to Dr. Katharina Dalton's claim that using natural progesterone from day eight through the end of her cycle will make a woman contraceptively safe, he stressed, *"Don't count on it.* I can think of three cases where it did not work, and they have three beautiful babies to show for it." (Dr. Lee informs us that Dr. Dalton recommends much larger doses than most doctors.)

Wallace Simons added, "We have women who are using synthetic progestogen pills for birth control and they absolutely will not give up this protection, yet want to use the natural progesterone for their PMS problems and especially to ward off the side effects of birth control pills. At first we said, 'Sorry, you must get off the progestogens first.'" But the pharmacy had so many requests for this combination that they finally began to recommend the natural progesterone along with the synthetic after the tenth pill of the pill pack during the second half of the menstrual cycle, and it worked just beautifully.

So even though the subjects' receptor cells have been at least partially blocked by the synthetics, the body somehow utilizes the natural progesterone to offset any discomfort. I asked Simons if there was any way one could use just the natural progesterone for ovulation control. He replied, "No, and I have a two-year-old granddaughter to prove that. We cannot tell you it's a birth control."

In essence, then, the authorities seem to agree on the principle; it's the application that can be risky. For greater safety it might be wise to combine natural progesterone for ovulation control with a barrier method of birth control. As for surgical contraception, too few women are aware of Dr.

Dalton's warning: "Sterilization should be avoided; it can lower the blood level of progesterone and increase PMS."[105]

FERTILITY: CARRYING AN INFANT FULL-TERM

Increasingly in the area of fertility therapy, medical doctors are concentrating on prescribing more natural substances for women. It would be wise to try to locate one in your area who will work with you and understand your needs when it comes to natural hormone replacement therapy.* A case in point: natural progesterone for conception. Although under ideal conditions it sometimes works as a contraceptive, progesterone is also, conversely, used in some fertility clinics.[106]

Jerome Check, M.D., an infertility specialist and professor of obstetrics and gynecology at Thomas Jefferson University and Hahnemann University, says that "too often physicians will treat the infertility problem with strong medication or even surgery without checking progesterone levels first. . . . But for many women, progesterone therapy has been very effective in helping them to become pregnant and to carry the child to term. Only after this treatment is tried, should more drastic procedures be considered."[107]

An adequate amount of progesterone is crucial to a woman who is trying to become pregnant. It actually prepares the uterine wall for implantation of the fertilized egg. Without sufficient progesterone, the egg will be expelled. Progesterone treatment can also be used to induce fertility when there appears to be ovulatory dysfunction (most commonly a luteal phase defect).[108] A study was performed involving fifty women who had lived with infertility for a minimum of one and a half years. Seventy percent of the women conceived within six months while exclusively using progesterone therapy, reports Dr. Check. "The Efficacy of Progesterone in Achieving Successful Pregnancy" describes this group:

> Five patients had a history of previous spontaneous abortions; all others had primary infertility. The range of ages was 18 to 39, with an average of 31. Their average period of infertility was 2.8 years in the 35 patients who conceived, and 2.7 years for the entire group.[109]

From all the data it seems clear that natural progesterone therapy offers no risks to the patient and will be likely to benefit those wishing to

---

* Information for your physician can be found in appendix F, and pharmacies and distributors of natural HRT are listed in appendix G.

conceive. Additional reports indicate that without progesterone treatment, women with luteal phase defect are at very high risk for spontaneous abortion. Progesterone has been found to be important in maintaining a pregnancy during the early months.

Several other studies regarding estriol's role in promoting fertility have been published. The clinical trials have shown that taking 1 to 2 mg of estriol daily on days 6 to 15 of the cycle seems to be sufficient to stimulate an increase in fertility in most cases. Doses of up to 8 mg of estriol daily may be used without adversely affecting cycle length.[110] Estriol improves fertility by improving the quality and quantity of the cervical mucosa. In a study of ten infertile women, 40 percent became pregnant after using estriol.[111] As always, the addition of progesterone is recommended with any estriol treatment (see chapter 2 for details).

You can learn to chart your own most fertile time of month by your basal body temperature and changes in your saliva and cervical mucus. By keeping records in the form of a graph, you will learn your ovulation pattern. Whether you are trying to conceive or to avoid conception, this information will be of value to you.

If you take your temperature every morning before rising, you will notice that it dips and then shoots up around the time of ovulation. Likewise, your vaginal fluid will become more slippery, wet, and copious at this time, with a consistency and color similar to egg white.[112]

Last, your saliva, when dry, forms a unique crystalline pattern during a seven- to ten-day period around mid-month. Ovulation is also sometimes signaled by spotting, a twinge of abdominal pain, or breast tenderness. For more information on saliva testing or other ovulation signs, contact the following:

Light on Earth Products at (415) 389-5400 will provide you with a microscope for home use so that you can track your own fertility patterns. An excellent aid for self-testing of saliva or cervical mucus, this lens will accurately pinpoint the fertile period, identify hormonal imbalance relating to infertility, and check for ovulation during lactation and premenopause. Along with the scope comes a tracking chart and an instructional booklet, which will alert you to possible problems such as miscarriage.

Body Wisdom: My Body Fertility Awareness System at (800) 888-9897 offers a fertility monitoring kit, a 35-minute video, a basal body thermometer, a booklet with tracking charts, and a pocket microscope for detecting the rise of estrogen.

PREGNANCY AND PROGESTERONE

A study reported in the *British Journal of Psychiatry* observed that administering progesterone from the middle trimester of pregnancy for relief of the symptoms of toxemia had some unexpected benefits: "A significant improvement in educational performance was demonstrated among children [whose mothers] received progesterone before the sixteenth week" following conception; and after giving birth their mothers seemed to have greater success at breastfeeding.[113] Clinical observations involving ninety children whose mothers received progesterone were summarized thus:

> More progesterone children were breast-fed at six months, more were standing and walking at one year, and at the age of 9–10 years the progesterone children received significantly better gradings than controls in academic subjects, verbal reasoning, English, arithmetic, [and] craftwork, but showed only average gradings in physical education.[114]

Dr. Katharina Dalton, who conducted these studies, first discovered the amazing benefits of progesterone through personal experience when she found that her own menstrual migraines disappeared during the last six months of pregnancy. She concluded that the high levels of progesterone during pregnancy might have made the difference. She then tested the use of progesterone on other women and found the same rapid relief of both headaches and other symptoms. Noting that if symptoms normally associated with PMS should return at any stage of pregnancy, a resumption of progesterone treatment would be indicated, she advises: "You could be wise to arrange prophylactic progesterone during pregnancy."[115]

Dr. Dalton is one of many scientists and doctors who have discovered that progesterone in the natural form

- protects the fetus from miscarriage;
- increases the feeling of well-being of the mother;
- increases the potential IQ of the child; and
- produces calmer, less colicky babies.[116]

To protect the fetus the body secretes ten to fifteen times more progesterone during pregnancy than at other times. Dr. Lee tells us that the placenta becomes the major source of progesterone, producing 300 to 400 mg per day during the third trimester.[117] What a great protection we have during pregnancy with this incredible hormone![118] And with no known dangerous side effects.[119–121]

Dr. Dalton calls morning sickness "a sign that the ovarian progesterone is insufficient and the placenta is not yet secreting enough progesterone." She says that giving the woman extra progesterone will ease the symptoms.[122] Also, the medical archives of the year 1952 include a study of seventy pregnant women with nausea and vomiting. After receiving additional supplementation of vitamins B$_6$, C, and K, 91 percent of these women found complete relief of their symptoms. Says Dr. Alan Gaby, "Although this study was reported in the *American Journal of Obstetrics and Gynecology*, very few obstetricians today know about it."[123] Many herbalists have also found ginger or cloves to be quite effective and safe remedies.

Biologist Margie Profet of Seattle has proposed a theory that morning sickness evolved as a way of protecting the fetus during its first, most vulnerable stage of development. The diet of early humans often included substantial amounts of wild plants that contained strong toxins as part of their own built-in defense system. While the adult woman may have developed a tolerance for these toxins, they might have been dangerous to the fetus. Thus, it's intriguing to postulate that nausea during early pregnancy discouraged a woman from eating very much, or perhaps interfered with her keeping foods down. She thereby would avoid ingesting (or retaining) poisons and consequently passing them on to her developing child. This may have brought about "morning sickness" as a protective adaptation,[124] triggered by the hormonal changes of pregnancy.

Dr. Ray Peat adds, "Since natural progesterone has been found to reduce the incidence of birth defects, it would seem reasonable to be sure that your own progesterone has returned to normal before getting pregnant."[125]

MISCARRIAGES

If a woman has had four or five miscarriages in the first six or eight weeks of a pregnancy, this is always due to luteal phase failure, says Dr. John Lee. Progesterone is needed to facilitate implantation and to prevent rejection of the developing embryo, but the follicle may not respond to the ovum with enough. Dr. Lee's recommendation: "Wait till you ovulate, and then four to six days after possible conception do a blood test (for HCG) to see if you're pregnant. If you are, start the progesterone; that way you will increase your chance of having a healthy baby." Blood tests for pregnancy tend to be positive within seventy-two hours of conception, whereas he says urine pregnancy tests are not usually positive until two weeks after conception.

One of Dr. Lee's notable findings is that there is an immune-suppressing effect in the uterus from higher doses of progesterone. This is important, because when conception takes place, half of the baby's chromosomes are from the male and half of them from the female. That makes the baby's tissue DNA different from the mother's because of the contribution of the father. If there's not a good tissue match, the difference will create tissue rejection. If you try to do a skin graft or a kidney or heart transplant and the tissue isn't the same, the body will reject it. But this doesn't happen with pregnancy. Why? Because of the progesterone response in the uterus. It's a site-selective action that doesn't occur anywhere else in the body; therefore, the baby is not rejected. By giving more progesterone after conception, you thus increase the likelihood that the baby will survive.

Looking at the problem from another perspective, Dr. Lita Lee informs us that "after conception progesterone prevents miscarriages resulting from excess estrogen."[126] It is interesting to note the consistency of the research, as in Dr. Peat's study, indicating that "pregnancy toxemia and tendency to miscarry or deliver prematurely are often corrected by progesterone."[127–129] Dr. Peat goes on to say, "My dissertation research, which established that an estrogen excess kills the embryo by suffocation, and that progesterone protects the embryo by promoting the delivery of both oxygen and glucose, didn't strike a responsive chord in the journals which are heavily influenced by funds from the drug industry."[130]

It is a fact that if a pregnant woman produces too much estrogen, her embryo can be suffocated (hypoxia). Dr. Lita Lee cautions that during the ninth week of pregnancy, a woman can lose her baby if she is a "high estrogen producer and/or [is] consuming commercial meat, poultry and dairy products containing synthetic estrogen (DES)." However, she goes on to say that natural progesterone "has been known to protect against the toxic effects of excess estrogen, including abortion."[131] Make certain, if hormones are prescribed during pregnancy, that they are not the synthetic progestins or estrogens but the natural micronized products. We now know that artificial hormones can be dangerous to the fetus during pregnancy.[132]

Dr. John Lee stresses that synthetic compounds cannot be efficiently "excreted by one's usual enzymatic mechanisms. Despite their advertisements, synthetic hormones are not equivalent to natural hormones."[133] Side effects can include fatigue, elevation of cholesterol, heart palpitations, headaches, depression, emotional disorders, weight gain, bloating, and more.[134]

MOTHERHOOD AND POSTNATAL DEPRESSION

After the birth of a child, symptoms may develop in the mother that she cannot control—and supplementation with natural progesterone is not considered a therapeutic option by mainstream medical practice. However, if you have depression or insomnia, find that you want to sleep all the time, or have mood swings or any of the signs we've mentioned in the introductory chapter of this book, think twice and have your doctor test you for progesterone deficiency.

This deficiency may be the culprit if you are suffering from anxiety or "the blues" after the birth of your baby. Many women feel they have to resort to medication or a support group for postpartum depression. However, the preventive step of taking natural progesterone immediately after childbirth might reduce their need for such measures. Dr. Dalton explains why:

> At delivery of the baby the placenta [which produces progesterone] is also delivered, and there is a sudden alteration in the levels of all hormones. The new mother must abruptly adjust to the complete absence of progesterone after nine months of a continuous and plentiful supply. It is suggested that some women find this alteration of progesterone level difficult to tolerate and react with the development of postnatal depression.[135]

It is very possible that temporary supplementation with progesterone will prevent anxiety and uncomfortable physical problems, as indicated in Dr. Dalton's *Guide to Progesterone for Postnatal Depression*:

> Progesterone is best used . . . at the completion of labor to prevent postnatal depression. . . . Progesterone enhances lactation. . . . Progesterone therapy may be combined with the patient's usual medication, be it antidepressants, tranquilizers, beta blockers, anticonvulsants etc. When symptoms abate the other medication can be gradually reduced.[136]

She continues with this caution about prescribing synthetic birth control pills to such women: "Progestogens . . . are man-made oral substitutes for progesterone, and are present in all oral contraceptives. They are known to lower the blood progesterone level. Thus oral contraceptives should not be used in any women at risk of, or currently suffering from, postnatal depression."[137]

Nursing Mothers

Many women join LaLeche League, a support group that helps mothers who want to nurse their babies and gain information and confidence about keeping their infants healthy. Seldom are they told, however, that in excess, "estrogens may reduce the flow of breast milk," as reported in *The Pill Book*.[138] One of the reasons so many women become discouraged and switch to bottle formulas is the frustration of not being able to produce enough breast milk. Dr. Dalton mentions extensive studies that link mammary development to the amount of progesterone prescribed to mothers during pregnancy. It was found that the women who had been given natural progesterone were able to sustain breastfeeding of their infants for at least six months after birth. Furthermore, the likelihood of success at breastfeeding was related to the dosage of progesterone administered.[139] (For mothers who, for one reason or another, are not able to breastfeed, see information on healthy formulas in chapter 6 and appendix C.)

Some Closing Thoughts

The better I've begun to feel, the greater has been my desire to learn why, and to share my findings with all who would listen. The more I've read, the more excited I've become to be part of a generation that is making such progress in understanding the impact of natural medicine. I realize that if we spread the word concerning the powerful and wonderful gifts of nature, perhaps others will benefit physically, mentally, and even spiritually from our newfound freedom of choice.

CHAPTER 4

CAN WE CIRCUMVENT OSTEOPOROSIS?

*There is one thing stronger than all the armies in the world,
and that is an idea whose time has come.*

Victor Hugo[1]

It was an unsettling revelation to me that osteoporosis can begin as early as fifteen years prior to the first signs of menopause—often around the middle to late thirties.[2] By the time most women reach their postmenopausal years, the majority will suffer from this disease—a fact that has made it the most common metabolic bone disease in this country.

The gradual loss of bone, perhaps 1 percent each year at first, accelerates to a rate of 3 to 5 percent per year during menopause and then reverts to about 1 to 1.5 percent a year thereafter.[3] This association of accelerated bone loss with menopause, first recognized over fifty years ago, led medical doctors to prescribe estrogen supplements during menopause to reduce these chances. Unfortunately, however, there are some problems with this approach. Of great importance are the significant side effects that start appearing in a woman's body when supplemental estrogen, unopposed by natural progesterone, is introduced. They constitute a long list, ranging from increased blood clotting and water retention to liver dysfunction and greater risk of endometrial and breast cancer.

As if that weren't bad enough, it also turns out that this estrogen therapy doesn't really do very much good. Nevertheless, the standard medical wisdom continues to support this approach and to assume that it is the most effective treatment. There is ample evidence in the medical literature that the therapy is of some limited value, at best, during the menopausal years. However, according to Sandra Cabot, M.D., "when estrogen is discontinued, calcium loss resumes."[4] So we need to look much more closely at the conventional method of treatment.

Dr. John Lee suggests instead that this escalating bone loss is due to decreasing levels of progesterone, caused by failure to ovulate during some menstrual cycles—for progesterone is mainly produced in the process of ovulation. In nonpregnant ovulating women, the ovaries normally produce 20 to 40 mg of progesterone daily during the second half of the menstrual cycle.[5] During pregnancy the placenta becomes the main producer of progesterone, making ever-increasing amounts, so that by the last three months of pregnancy, it is making 300 to 400 mg a day. Failure to produce these levels of progesterone naturally can lead to trouble. Even though estrogen aids in slowing down bone loss, progesterone could be called proactive, since its stimulatory effect on the osteoblastic cells actually encourages bone growth.[6]

THE IMPORTANCE OF OVULATION

The onset of irregular periods is an indicator that progesterone levels are becoming depleted with respect to estrogen. When menopause is upon us (that is, when we have stopped ovulating), our blood progesterone will decline to almost zero.[7] A reasonable question would be, "Why do some women experience this sooner than others?" Researchers tell us that stress, injury, poor diet, lack of exercise, and trauma all may play a role in the degree to which ovulation becomes sporadic and then tapers off at menopause.[8]

To these Dr. John Lee would add the damage done to the ovaries by any of the many human-made estrogenic chemicals in the environment (see chapter 5). Such exposure to the female fetus or very early in life may damage the ovarian follicles to the extent that in adulthood they can no longer make progesterone as they should. Follicle dysfunction induced by these so-called xenoestrogens may well be the primary cause of the progesterone deficiency that often occurs fifteen years or more before actual menopause.

In addition, as is widely reported in the press these days, the way you treat your body in general can contribute to premature bone loss. Smoking and excessive consumption of alcohol, caffeine, soft drinks, and meat protein, as well as the use of certain anti-inflammatory or antiseizure medications or thyroid hormone replacements, may all place you at higher risk. And some factors can't be avoided: thin, small-boned women and those of Caucasian descent have a higher risk of osteoporosis.[9]

In the United States, approximately 24 million people are affected by osteoporosis, at a medical cost of over $10 billion and as many as 1.5 million fractures[10] leading to disability, deterioration, and, for too many, death. Today, the annual number of fractures attributable to osteoporosis continues to escalate as our exposure to estrogen from various sources has drastically increased.[11] But, as Dr. Robert Lindsay has said, "The problem is, nobody feels the bone they're losing until it's too late. . . . Osteoporosis is without symptoms until it becomes disease."[12] According to Dr. Patricia Allen, when the "acceleration of bone loss begins, risks for coronary artery disease start to increase [and] atrophy of breast and genital tissue starts. And so most doctors now believe that a woman who is bothered by menopausal symptoms should be treated before the cessation of her periods."[13]

PROGESTERONE (NOT ESTROGEN) SUPPLEMENTS FOR HEALTHY BONES

Jerilynn C. Prior, M.D., and her associates found evidence of progesterone's possible role in countering osteoporosis in a study of sixty-six premeno-pausal women between twenty-one and forty-one years of age. All these women were long-distance marathon runners. It was observed after twelve months that

> the average spinal bone density decreased by about 2%. . . . However, women who developed ovulation disturbances during the study lost 4.2% of their bone mass in one year. While there was no correlation between the rate of bone losses and serum levels of estrogen, there was a close relationship between indicators of progesterone status and bone loss.[14]

Now this is news! And then *Medical Hypotheses* claims that the use of natural progesterone is not only safer but less expensive than using its synthetic formulation, Provera (medroxyprogesterone), and that *"progester-one and not estrogen is the missing factor . . . in reversing osteoporosis."*
The journal continues:

> The presence or absence of estrogen supplements had no discernible effect on osteoporosis benefits. . . . Progesterone deficiency rather than estrogen deficiency is a major factor in the pathogenesis of menopausal osteoporosis. Other factors promoting osteoporosis are excess protein intake, lack of exercise, cigarette smoking, and inad-equate vitamins A, D, and C.[15]

Dr. Majid Ali says that the use of estrogen to prevent osteoporosis is really quite "frivolous."[16] Osteoporosis is a disease we can do much to prevent. With the knowledge we presently have, it is imperative that women take active steps toward a healthier lifestyle. We must take to heart what author Gail Sheehy says in *The Silent Passage: Menopause*:

> Nearly half of all people over age seventy-five will be affected by porous bones causing the risk of fractures of many kinds. The National Osteoporosis Foundation in the U.S. says that almost a third of women aged sixty-five and over will suffer spinal fractures. And of those who fall and fracture a hip, one in five will not survive a year (usually because of postsurgical complications).[17]

It has been estimated that twice as many serious fractures occur today than thirty years ago. How long will it take us to grasp the truth of the matter, so we can help ourselves and the aging population? "Clearly," says Dr. Alan Gaby, "there is something wrong with our bone health, something that the medical profession has not been able to do much about. There is more to preventing bone loss than calcium supplements, estrogen replacement therapy and exercise."[18]

These reminders about the decline in bone mass as we age make me think of my own family gatherings during the holidays, when we are at long last in the company of several generations of family members. Someone usually says, "Haven't you grown!" In our family we take it a step further: someone stands next to Mom, and then Mom next to Grandma—and, sure enough, there is a definite change! But it's in the opposite direction. Soon a grandchild will say, "Wait a minute, Grandma, aren't you shrinking?" It seems that these changes start earlier than we may think and are more crippling than we realize.

Is this a topic we can continue to take lightly? Not according to Robert P. Heaney, M.D., professor of medicine at Creighton University School of Medicine in Nebraska. In commenting on the medical community's having overlooked the importance of progesterone in osteoporosis, he expressed the hope that research will "galvanize the field into taking the matter seriously."[19] Perhaps statements such as his will begin to reeducate the very doctors who think they know all there is to know about this most vital subject.

It is a mystery that so much focus has been placed on declining estrogen levels; it seems the emphasis has been on the wrong hormone. The October 14, 1993, issue of the *New England Journal of Medicine* makes it clear that taking estrogen for five or ten years after menopause will not protect a woman from having a hip fracture in her later years.[20] Why should we wait ten to twenty years for the results of the studies that are now in progress? We have already been counseled by many medical experts. Now is the time to make the change from an estrogen replacement program to one based on natural progesterone therapy.

We should ask ourselves, "Why would we use a hormone that has not worked for generations past?" The traditional and often one-sided references to estrogen decline have created a body of misinformation that has sentenced many to poor health and needless distress. It seems irresponsible that the medical world is not doing double-blind studies, along with baseline and follow-up bone mineral density tests, with natural progesterone.

However, we can be grateful to the many doctors who have searched the archives for the truth of the matter. We now have reliable evidence that despite declining estrogen levels, bone loss accelerates when progesterone levels decline, and bone minerals can be restored with natural progesterone replacement therapy.[21] Yet, the message women receive from their medical doctors is that "estrogen is the single most potent factor in prevention of bone loss."[22] This belief has been handed down from one generation to the next. Fortunately, recently published studies and books are now challenging the medical theory and bringing more light to the subject of preventing osteoporosis.

A case in point is the book *Preventing and Reversing Osteoporosis*, written by medical doctor Alan Gaby. I became so absorbed in it that I could not put it down—nor will you, when you find that, yes, osteoporosis *can be reversed*. Much of what Dr. Gaby says would be beneficial to many and should be shared. He cautions that despite the preventive measures of calcium supplementation and exercise, and despite medical intervention with estrogen therapy, osteoporosis is getting worse: "At least 1.2 million women suffer fractures each year as a direct result of osteoporosis. . . . Fractures seem to be increasing, . . . and this difference cannot be explained by the aging of the population."[23]

Let us hope that more medical doctors are getting away from the mainstream of drug therapy and are discovering natural remedies that seem to work more efficiently for such problems in the long run. Dr. Gaby, for instance, with twenty years of medical research and thirteen years of clinical practice, writes that many of the most significant advances and effective treatments have been those discovered or administered outside the auspices of the traditional medical community.

Dr. John Lee comments that modern medicine "strangely persists in the single-minded belief that estrogen is the mainstay of osteoporosis treatment for women." Strange, indeed, that doctors should think like this, when even medical textbooks such as *Harrison's Principles of Internal*

Medicine (12th edition, 1991) and *Cecil's Textbook of Medicine* (18th edition, 1988) don't back up this theory.[24] Along the same lines, Dr. Lee also quotes the 1991 *Scientific American Medicine:*

> "Estrogens decrease bone resorption" but "associated with the decrease in bone resorption is a decrease in bone formation. Therefore, estrogens should not be expected to increase bone mass." The authors also discuss estrogen side effects including the risk of endometrial cancer which "is increased six-fold in women who receive estrogen therapy for up to five years; the risk is increased 15-fold in long-term users."[25]

PROGESTERONE CREAM FOR OSTEOPOROTIC PATIENTS

Although there are many forms and ways to take natural progesterone, Dr. Lee acquaints us with the transdermal method. By carefully observing his patients over the course of fifteen years, he proved the effectiveness of transdermal progesterone cream. His work confirmed its safety and its remarkable benefits to his osteoporotic patients who had a history of cancer of the uterus or breast and to those who had diabetes, vascular disorders, and other conditions.

Dr. Lee had hoped that the progesterone would strengthen his patients' bones. To his surprise, it did; their bone mineral density tests showed progressive improvement and the number of his patients suffering osteoporotic fracture dropped nearly to zero.[26]

Dr. Lee is perplexed at "the reluctance of contemporary medicine to adopt the use of natural progesterone." It's his impression, however, that "the news is spreading and change is on the way." In the publication *Natural Solutions*, Dr. Lee voices true dismay with his orthopedic colleagues who chose not to use the progesterone cream in their patients' care "but did put their own wives on the cream."[27]

Dr. Lee points out that the "conventional treatment of osteoporosis with estrogen, with or without supplemental calcium and vitamin D, tends to delay bone mass loss, but not reverse it."[28] His investigation into using transdermal progesterone cream instead of a synthetic estrogen replacement treatment demonstrates that "osteoporosis subsided, musculoskeletal strength and mobility increased, and monthly vaginal bleeding did not occur."[29] Most striking were the results of the dual-photon densitometry test, which measured a 10 to 15 percent increase in bone mineral density, even in women who had experienced menopause twenty-five years earlier.[30]

After years of researching transdermal progesterone supplementation,

Dr. Lee observed in his patients "a progressive increase in bone mineral density and definite clinical improvement including fracture prevention. . . ." He concluded that "osteoporosis reversal is a clinical reality using a natural form of progesterone derived from yams that is safe, uncomplicated and inexpensive."[31] Unfortunately, by the time many of us are ready to deal with the impact of osteoporosis it has already done considerable damage, as it is symptomless until the fractures begin. If you think that you can deal with brittle bones after you get through the inconvenience of hot flashes and night sweats, you need to think again.

It is an enigma to me that our nation's supposedly up-to-date medical researchers continue to be oblivious to the evidence that progesterone stimulates new bone formation by the osteoblasts, the bone-building cells.[32] Think of the many aging women who could benefit from this information and be freed from unnecessary pain and spared their disability. As Gail Sheehy observes, osteoporosis "often leaves older women frail, susceptible to falls and broken bones. . . . [It] makes it painful merely to sit." Many elderly osteoporotic women die of secondary infections following hip surgery.[33] These infections are what makes osteoporosis victims subject to death, not the osteoporosis itself.

Reading about this reminded me again of my mother's fragile condition as her hip bones grew so weak she could hardly get out of a chair. The longer she sat in one place, the more pain she felt. Before long she had to depend on a wheelchair to get around, and in an even shorter time she yielded to a hospital bed in our home. We felt blessed that she at least did not have to enter a nursing home, as so many do.

BONE MINERAL DENSITY TESTS

Establishing a baseline bone mineral density (BMD) is of vital importance for all women as they make the transition from monthly ovulatory cycles to the menopausal stage that heralds the onset of accelerated bone loss. Currently, the most precise test with the least radiation dose is dual X-ray absorptiometry (DEXA or DXA), which yields an accuracy of 98 to 99 percent on a study of the lumbar spine[34] and hips. Another highly reliable scan is the dual photon absorptiometry (DPA) test, which is 95 to 98 percent accurate and uses photons (high-intensity light beams) to determine the density of your bone. The importance of such testing cannot be overemphasized, as osteoporosis is asymptomatic in its earlier stages. Multiple scans, several years apart, from the mid-forties (baseline evaluation) to the mid-sixties will give women an accurate picture of their BMD status as well as feedback regarding the efficacy of the HRT program they are using. (Please see chapter 7 for additional diagnostic resources.)

Dr. Lee has found these tests extremely useful. One woman in her seventies consulted him for her advanced osteoporosis and spinal compression fractures:

> She had previously avoided hormone therapy because of a long history of fibrocystic breasts prior to menopause. With natural progesterone applications, her BMD (bone mineral density test) rose gratifyingly, her back pains disappeared, and she resumed normal activities such as hiking, boating, gardening, etc.[35]

The women who received the best results from the progesterone studies were those who seemed to need it the most. In other words, as progesterone was administered to women with all levels of bone loss, regardless of age (whether seventy or thirty-five), those with the lowest initial bone densities had the greatest increases in bone mass.[36]

Dr. Lee cautions against relying on the hair analysis some laboratories do to supposedly diagnose osteoporosis. The reason is that at this point in the resorption process, calcium will naturally register high—because all the calcium that has been released from the bone is now circulating in the bloodstream and is being picked up by the hair. When the lab tells you you have a good level of calcium and there is no need to worry about osteoporosis, be wary; he says it is wrong to use this test to measure whether or not osteoporosis is present and extra magnesium and calcium are needed.

HOW OLD IS TOO OLD FOR PROGESTERONE?

A study was conducted with a hundred patients who ranged in age from thirty-eight to eighty-three years. They were all menopausal or postmenopausal. The majority of these women had already noticed a loss of height due to compression fracture of the spinal vertebrae from age-related bone thinning. Dr. Lee calls this "a cardinal sign of osteoporosis." A number had also experienced fractures of other skeletal bones, such as hips or ribs. We read in *Medical Hypotheses:*

> Since the U.S. medical insurance does not include payment for dual

photon bone density tests (approximately $150/test),* only 63 of the patients were included to submit to serial testing. Thus, 37 patients could be followed only by clinical signs, i.e., relief of osteoporotic symptoms and reduction in expected fracture incidence. Even so, the benefits from the treatment program were so obvious to these patients that no problems with patient compliance arose. No side-effects or adverse alterations in blood lipids were observed. Each patient was followed a minimum of 3 years.[37]

Of the hundred patients in this group, the report states that "height loss was stabilized" and there was an associated relief from osteoporotic pains. The amazing findings of the test given to determine bone density measurements of the lumbar vertebrae (serial DPA) included (1) reduction in fracture incidence and (2) increase in bone density. Most of these women were responding to a transdermal progesterone dose of only 240 mg per month or 10 mg per day. In the sixty-three women given natural progesterone, the benefit proved to be extraordinary—showing that in three years, instead of losing 4.5 percent of bone as expected, the subjects actually increased their bone density by 15.4 percent, regardless of their age. This report confirms that the greatest relative improvement was made by those with the lowest bone density to begin with.[38]

Further, the study refuted the myth that osteoporosis is irreversible for older women, or even that it is more difficult to correct. Indeed, test subjects over the age of seventy (the oldest was eighty-three) showed somewhat better results than those under the age of seventy. The former responded to the progesterone therapy regimen with an average increase in bone density over the three-year period of 14.4 percent, while the younger women showed an increase of just 14.0 percent.[39]

Many similar studies can be found in the *International Clinical Nutrition Review*. In them, not only were the benefits to skeletal strength observed, but patient complaints such as gastric irritation, joint stiffness, moodiness, and headaches were also relieved. Aside from the safety and effectiveness of using a natural progesterone, it has proven to be extremely cost-efficient—about one-tenth the cost of the same dosage of medroxyprogesterone (Provera), and with no side effects.[40]

---

* Some insurance plans now offer coverage for osteoporosis testing. Unfortunately, they may require your doctor to certify that you are estrogen-deficient, with no consideration of progesterone levels.

We need to reiterate, then, the important point that Dr. Lee makes in the *International Clinical Nutrition Review*. Natural transdermal progesterone cream, he says, "is the missing link in healthy bone building in postmenopausal women." Concerning osteoporosis he says, "Reversal has been demonstrated by the bone density tests and by the clinical results. This cannot be said of any other conventional therapy for osteoporosis."[41]

The conclusion of twelve different reports is that progesterone deficiency rather than estrogen deficiency is one of the main factors in the development of menopausal osteoporosis. Dr. Lee found the best overall results when his patients combined use of the natural progesterone with optimal nutrition. Women were given a "low protein, high vegetable diet, modest exercise and vitamin supplementation." Dr. Lee recommended supplemental estrogen in the form of estriol only when his patients experienced hot flashes, cystitis, or vaginal dryness that did not subside after two to three months of natural progesterone treatment.[42] (Also see appendix A.)

Dr. Lee informs us that "age is not the cause of osteoporosis; poor nutrition, lack of exercise, and progesterone deficiency are the major factors."[43] He says osteoporosis seems to be more common in "white women of northern European extraction who are relatively thin" or who smoke cigarettes, are under-exercised, are deficient in vitamins A, D, or C, calcium, or magnesium, or whose diet is meat-based rather than vegetable- and whole grain-based.[44] Dr. Gaby confirms some of these observations on the basis of numerous studies. It was found that the most comprehensive program (one that included proper diet, botanical progesterone, and a broad spectrum of vitamins and minerals) produced an astounding 11 percent increase in bone mineral content in postmenopausal women in less than one year. Any one aspect of treatment used alone could never have come close to bringing such improvement in such a short period of time.[45]

Jonathan V. Wright, M.D., an expert in nutritional biochemistry, applauds Dr. Gaby for going "beyond the calcium craze to a holistic approach to healthy bones." He says, "What's good for the bones is good for the heart, the skin, the breasts, the stomach, and even crucial for future generations."[46]

We need to take these experts' advice and question the end result of the extra protein we eat. "A nutritional program that is more than 30 percent protein increases calcium excretion," says Serafina Corsello, M.D., "because metabolism of the protein acidifies the body, which then tries to achieve normal alkaline balance by excreting calcium."[47]

MEAT PROTEIN, PHOSPHATES, AND BONE LOSS

Both kidney stones and kidney failure are an indication that acidosis may be silently weakening our bones as well. In the periodical *Health Science,* we learn that meat eaters have a much higher rate of osteoporosis than vegetarians, even though vegetarians have a lower calcium intake. The conclusion from this research is that the best prevention for osteoporosis is a relatively low-protein diet and plenty of fresh fruits and vegetables. This may provide enough calcium without the need for dairy products. In fact, who would believe that a half cup of sesame seeds contains 870 mg of calcium?[48]

According to Joel Fuhrman, M.D., protein from meat also contains a large quantity of disulfide bonds, which, "while undergoing oxidation when broken down, create sulfate and hydrogen ions that further increase the acid load in the blood. To neutralize this acid load, the body calls on its bony stores of calcium to provide basic [alkaline] calcium salts. . . . In addition, urea and other waste products from excess protein digestion cause the kidney to work harder and excrete more fluid and with this increased function more calcium is lost in the process."[49] Under these considerations the kidney is not able to reabsorb calcium before it is evacuated, and the consequence is more calcium lost via the urine.

Dr. Fuhrman cautions that calcium loss is stimulated in a variety of ways. For instance, with cigarette smoking, the nicotine disrupts hormone communication to the kidneys, curtailing calcium reabsorption. Antacids (with aluminum added), diuretics, and antibiotics also contribute to the loss.[50] And most significant, affecting young and old alike, are soft drinks. Read the labels before you quench your thirst. Many sodas contain not only caffeine but sodium and phosphoric acid—all of which contribute to bone loss in different ways.[51] As Dr. Fuhrman tells us in *Health Science:*

> Phosphates have been shown to increase the release of parathyroid hormones, which mobilizes skeletal calcium reserves. The phosphoric acid found in carbonated drinks is particularly damaging, and is more powerful in inducing calcium excretion than is the phosphorus contained in natural foods.[52]

People who habitually drink soda are oblivious to the fact that many brands contain a dangerous ingredient, phosphate. Unknown to most consumers, this substance is insidiously robbing calcium from our teeth and bones. Phosphates are even added to processed foods such as cheeses and meats. If soda drinkers also eat too much meat (high in phosphorus), they are adding to an already serious problem.[53] This aspect of malnutrition may be causing unsuspected degenerative problems in many thousands of Americans.

A sampling of foods relatively high in phosphorus includes processed meats, turkey, ham, pork, fried potatoes, and crackers.[54] *The Herb Quarterly* reminds us also to avoid the following to prevent osteoporosis: dairy products, coffee, alcohol, salt, and sugar; and that eating "a diet rich in dark-green leafy vegetables, nuts, seeds, tofu, [and] molasses"[55] will assist in reversing osteoporosis.

According to the *American Journal of Clinical Nutrition,* studies involving sixteen hundred women disclosed that those who "follow a vegetarian diet for at least 20 years have 18-percent less bone mineral by age 80, whereas meat eaters have 35-percent less bone mineral."[56] Other interesting statistics have appeared in many nutritional reviews, such as this one from the Center for Women's Health at Columbia-Presbyterian Medical Center in New York: Following low-fat diets and consuming more vegetables is associated with fewer female disorders identified with PMS or menopause.[57]

THE CALCIUM MYTH

Many American women are being told by their physicians that one of the most efficient and inexpensive ways to supplement their calcium intake to prevent osteoporosis is with ordinary antacid medications (such as Tums). These contain mainly calcium carbonate, which may be inexpensive but is also one of the forms most poorly absorbed by the body.[58] According to John Mills, Total Quality Manager at the Highland Laboratory (Mt. Angel, Oregon), if your stomach is not functioning normally the calcium carbonate in the antacid will not remain dissolved long enough to be absorbed in the intestine and then be conveyed to where it is needed in the body (bones, teeth, and muscles). It will pass through the intestines without being assimilated, contributing to constipation.[59]

Antacids in general may temporarily relieve symptoms of digestion, but they do more harm than good in the long run as they may "contain aluminum, silicone, sugar, and a long list of dyes and preservatives, none of which will help you and may even harm you," reports Dr. John Lee.[60] He warns us, regardless of what your pharmacist or doctor says, not to try to obtain extra calcium by taking antacid tablets, as their side effects may far outweigh any benefits you would gain from their use.[61]

Other well-intentioned measures in common practice are also suspect. John McDougall, M.D., notes cases in which people took calcium without understanding what else contributes to its absorption or loss. Consumption of too much protein can result in excessive calcium excretion and cause the body's calcium reserve in the bones to be at risk. "Experiments have shown that when subjects consumed 75 grams of protein daily, even with an intake as high as 1400 milligrams of calcium, more calcium was lost in the urine than was actually absorbed."[62] (A more usable form of calcium is discussed below.)

THE CALCIUM BOOSTERS

Bones are dependent upon much more than just calcium. In his book on preventing osteoporosis, Dr. Gaby explains in detail:

> Magnesium is necessary to promote normal bone mineralization; silicon, manganese, and vitamin C are also essential for proper formation of cartilage and other organic components of bone; vitamin K is needed to attract calcium to the bones. It plays a role in remodeling and repair; vitamin D is necessary for absorption of calcium from the diet; zinc and copper are involved in repair mechanisms, presumably including those that occur in bone.[63]

His findings also suggest that "boron has a powerful influence on the metabolism of calcium, magnesium and some hormones."[64] The importance of magnesium is easy to underestimate. Dr. Lee affirms the fact that we need magnesium along with calcium. He says, "If magnesium is deficient, calcium is less likely to become bone and more likely to appear as calcification of tendon insertion points . . . leading to tendinitis, bursitis, arthritis, and bone spurs."[65] It is advised by some that if we choose to take calcium supplements, we should take twice as much magnesium as calcium on a daily basis (approximately 800 mg of calcium and 1,500 mg of magnesium).[66] Dr. Gaby says that magnesium has been a forgotten mineral and that at least a one-to-one ratio of magnesium-to-calcium (if not two-to-one) would lead to stronger bones and less inappropriate calcium deposition.[67]

Dr. Guy Abraham, a gynecologist in Torrance, California, also reversed the calcium-to-magnesium formula in a trial of twenty-six postmenopausal women. They were given daily supplements of 600 mg of magnesium (oxide) and 500 mg of calcium (citrate). The women were also on a high-vegetable, low-protein diet and their supplements included C, B complex, D, zinc copper, manganese, boron, etc. After eight to nine months the women's bone mineral density increased 11 percent with these higher amounts of magnesium.[68]

The more studies we come across, the more we realize that calcium used alone can cause serious problems. Dr. Morton Walker, too, states in *The Chelation Way* that for calcium to be properly metabolized, the right amounts of magnesium, phosphorus, and vitamins D (the "sunshine vitamin"), C, and A must also be present. When properly absorbed, he says, calcium is the nutrient that "helps to overcome cramping in the legs and feet."[69] To prevent brittle bones, Dr. Carlton Fredricks recommends cod liver oil for its vitamin D. (On the other hand, adds Dr. Ray Peat, cod liver oil somewhat increases one's need for vitamin E.) Soy protein may also contribute to healthy bones because of the isoflavones that soybeans contain.[70] (See chapter 6.)

Referring to a study in *Acta Endocrinologica*, Dr. Gaby comments that the mineral zinc "enhances the biochemical actions of vitamin D, which is itself involved in calcium absorption and osteoporosis prevention. Because of its essential role in DNA and protein synthesis, zinc is required for the formation of osteoblasts and osteoclasts, as well as for the synthesis of various proteins found in bone tissue." In another investigation drawn from *Acta Medica Scandinavica*, Dr. Gaby tells us that "zinc levels were found to be low in the serum and bone of elderly individuals with osteoporosis. . . . The most efficiently absorbed types of zinc," he advises, "are zinc picolinate, citrate and chelated."[71]

Dr. Lee adds that "zinc is essential as a co-catalyst for enzymes which convert betacarotene to vitamin A within cells. This is especially important in building the collagen matrix of cartilage and bone. As with magnesium, zinc is one of the minerals lost in the 'refining' of grain. As a result, the typical American diet is deficient in zinc and modest supplementation (15–30 mg/day) is recommended."[72]

Other nutrients with an important role in calcium absorption include silica and pectin.[73] Found in unpeeled apples, citrus fruits, the cabbage and broccoli family, and many other fruits and vegetables, pectin transports calcium molecules to the large intestine for slow absorption into the body (as well as neutralization of potentially cancerous toxins). The authors of *The Calcium Connection*, Drs. Cedric and Frank Garland, also stress the importance of drinking plenty of water to help dissolve and absorb dietary or supplemental calcium.[74]

Concerning calcium and hormones, Gail Sheehy advises that just taking calcium is not enough and that exercise *by itself* is also ineffective in preventing bone loss.[75] The combination of weight-bearing exercise, proper diet, and the appropriate kind of calcium (see Misconceptions about Calcium and Diet, this chapter), along with natural hormone replacement therapy, has been shown to increase bone mass and decrease symptoms of insomnia and hot flashes. Citing a study from the Netherlands, Gail Sheehy also notes that "vitamin K has been found to inhibit the precipitous loss of calcium in postmenopausal women by up to 50 percent. . . . Dark green leafy vegetables like broccoli and Brussels sprouts are sources of vitamin K."[76]

So we might want to think twice before a "professional" directs us to calcium supplements alone and ponder the words of well-known author and nutritional consultant Nancy Appleton, Ph.D., alerting us to the possible harmful aspects of calcium supplementation. She says, "Excess calcium can be redistributed in the body and is often deposited in soft tissues, possibly causing arthritis, arteriosclerosis, glaucoma, kidney stones and other problems."[77]

Furthermore, says Ruth Sackman in *Cancer Forum*, "Fragmented supplements [e.g., calcium that has been separated from other natural components of food] can actually cause a deficiency of the very supplement that is being used because it robs the body's storage areas (bones, nails, muscles, hair, etc.) in order to reconstruct the natural complex found in nature. To avoid the risk of bone loss from calcium deprivation, why not use foods rich in calcium?" From raw almonds to dried beans to parsley, many valuable foods are identified by books on nutrition.[78] Dr. Lee, pointing out that "cows get the calcium for their bones and their milk from plants they eat," says that among the best sources of dietary calcium are fruits and, especially, broadleaf vegetables.[79]

HIGH RISK OF BONE AND SPINAL FRACTURES

It's important to understand what our bodies need, since it is currently predicted that one out of every three women will eventually suffer from low bone mass and structural degeneration leading to fractures.[80] Women face at least a 15 percent risk of hip fracture. The annual cost of this trauma, which could be reduced by preventive efforts, is estimated at $7.3 billion in the United States. We need to focus on what we can do to circumvent such statistics in our lives—to achieve a healthy state of mind and body so as to have the energy to pursue our creative passions and goals to their fullest.

These figures constitute a national tragedy, considering that many of these women were faithfully subjecting themselves to synthetic HRT (with all its side effects)—and to what end? Their hips fractured just like those of the women who chose not to use any estrogen at all. This will seem even more of a crime when you read the following discussion of the role progesterone plays in bone formation.

This report, dealing with the prevention of osteoporosis, comes from the *Canadian Journal of OB/Gyn & Women's Health Care*. The authors, of the Division of Endocrinology and Metabolism at the University of British Columbia in Vancouver, state:

> Progesterone acts on bone, even though estrogen activity is low or absent. Because progesterone appears to work on the osteoblast to increase bone formation, it would complement the actions of estrogen to decrease bone resorption.[81]

The authors go on to explain that progesterone fastens to receptors on the osteoblasts (the bone-building cells) and "increases the rate of bone remodeling."[82] It is interesting to note once again that the role of estrogen is to slow down bone loss, while natural progesterone actually promotes bone production. Dr. Lee concludes from several studies that (1) estrogen slows down the dissolving of bone by the "osteoclast" cells, (2) natural progesterone stimulates the formation of new bone, and (3) certain progestins may also cause the osteoblasts to create a limited amount of new bone.[83]

Keep in mind that some studies have shown that the synthetic progestins normally prescribed by our doctors may actually diminish our supply of natural progesterone.[84] Many of our gynecologists do realize the destructive effect these synthetic progestins have on the body. This may be a major factor in why they prescribe estrogen alone.

BONE LOSS FROM COMMON MEDICATIONS

I believe we must always be on guard against any over-the-counter or prescription drugs that speed the dissolving of bone—a process called resorption. Trien Susan Falmholtz, in her book *Change of Life*, speaks about various commonly prescribed drugs that lead to the dreaded condition of osteoporosis. These include "thyroid replacement drugs; heparin (an anti-coagulant); cortisone preparations (such as prednisone); aluminum-containing antacids; anti-convulsants; and the antibiotic drug tetracycline."[85]

At the University of Massachusetts, a research project centered on the use of levothyroxine for treating thyroid problems. In certain circumstances this commonly prescribed hormone was found to cause as much as 13 percent bone loss.[86] A thyroid specialist I spoke with, however, says that this report has been quoted out of context and blown out of proportion. The main point to keep in mind is that dosages should never be prescribed in excess of demonstrated need, or for the wrong reason (such as for weight loss, when the gain was not caused by a thyroid deficiency).

Another example is giving thyroid medication simply for a low basal body temperature (a test often used by adherents of Dr. Broda Barnes), without other clear evidence of thyroid insufficiency. This type of test is also used for fertility problems to determine the time of ovulation and thus reflects a woman's progesterone level. Progesterone and thyroid are very much interrelated, but in some cases all that is needed is natural progesterone to correct an abnormally low body temperature. Dr. Peat believes that the mechanism by which progesterone raises the temperature prevents estrogen from blocking the action of thyroid hormone, because only in women is progesterone thermogenic.[87]

Dr. John Lee makes the point that a doctor who does not understand progesterone deficiency may give thyroid medication to someone who doesn't need it. This excess thyroid will increase the rate of bone resorption. However, when thyroid hormone is replaced for the purpose of restoring it to normal, it doesn't cause bone loss. In fact, the right dosage will improve bone density. As Dr. Lee explained to me, "It's not the thyroid that should be feared; it's the doctor who gives thyroid when you don't need it."

Noting another medicinal hazard, Dr. Alan Gaby describes a study of twenty patients who were scheduled for brain surgery. Some of them were given Maalox 70 to prevent stress and ulcers. This antacid is rich in aluminum, and the research team found, after analyzing brain and bone tissue, that this toxic substance had been absorbed into the body and deposited in the tissues. Dr. Gaby recapitulates information from the studies, saying that "the accumulation of aluminum in bone appears to reduce the formation of osteoid [areas of new bone], while at the same time increasing the amount of bone resorption. The result of this dual action of aluminum would be to accelerate bone loss."[88]

This is a good place to say just a word about the dangers of taking the latest "miracle" drug to prevent calcium loss. We are always reading or hearing in the media about experimental drugs, new hormones, or even new compounds that activate bone growth. Highly advertised drugs come onto the market quickly and leave with just as much speed. Rather than chase after the newest pharmaceutical product, we would do better to educate ourselves about the natural forms of healing that are not yet as widely promoted. It may take an assertive plan coupled with intelligent awareness to guard ourselves and guide us toward what is truly good for the healing and building of bones.

In exploring the importance of enzymes and vitamins to aid in the digestion of meat protein so that calcium will not be lost in the blood, I came across some information regarding salmon calcitonin that at first reading seemed to make it another good choice for those who are suffering from severe bone loss.

Few physicians seem to be aware of salmon calcitonin, which slows calcium resorption. Calcitonin is "found in humans (as well as in salmon, pigs, eels and sheep),"[89] according to an article in *Prevention* magazine. The author, Sydney Bonnick, says that calcitonin is a hormone that is manufactured by the thyroid gland. Its function is to deter the movement of calcium from the bone to the blood. Calcitonin has even been found to "relieve pain from fractured bones, and [it] may actually stimulate the formation of new bone."[90]

The reason it is not used more readily for those suffering from osteoporosis is that it is extremely expensive, costing approximately $2,700 per year; and it is quite inconvenient to administer via intramuscular or subcutaneous injections.[91] In the United States it is currently delivered by means of a daily injection of 100 I.U.[92] Although the synthetic versions of calcitonin, such as Calcimar, have been approved by the FDA for the treatment of osteoporosis, the literature on these products discusses possible severe allergic reactions, including anaphylactic shock.[93]

VOLUNTEER STUDY GROUPS

The above studies should make us more cautious when we hear about osteoporosis study groups that provide free medication. The other day such an experiment was announced on the radio, touting a miraculous "break-through in medicine" that sounded like the wonder drug of the century for the prevention of osteoporosis.

Curiosity got the better of me; being most interested in the subject, I had to call the toll-free number and ask the name of the drug. The person who answered said, "You are probably well aware that estrogen is needed for all women suffering from osteoporosis." She went on to say that estrogen was essential for the building of bones, and without its use we will increase our chances of osteoporosis. She then invited me to become part of this study group. I asked her whether the estrogen was natural. She said, "It is a synthetic estrogen called Ralaxifene." Thus ended my curiosity—but not my desire to want to reach those who would say "yes" to this proposal.

All of this points up the potentially catastrophic consequences of not being informed about natural health care. Not long ago I met a woman in my workplace who must have been in her sixties. She was walking with a lot of difficulty. I was telling her what I'd learned about hormones and osteoporosis. She said to me that she had always refused to take pills of any kind; in fact, she was quite vehement about this. She told me that ten years earlier she'd had a hip replacement, and because her other hip was now showing the same signs and giving her trouble, she had been asked to be a candidate for a national double-blind osteoporosis study group.

I felt so sad for her, because she'd been so determined to stay healthy by not taking any medicine—and now here she was, desperately taking pills every morning and evening. She was not even allowed to know whether she was taking a placebo or an active substance. She was living day to day with little hope—only the fear that her other hip would also need to be replaced. She looked at me and said, "What you're saying about hormone deficiency is interesting, because all my problems started when I was in my forties, right after my hysterectomy."

I thought to myself, "For twenty years she has had no hormone replacement therapy." This woman, prior to her hysterectomy, had been very active and pain-free, and I could tell she had so much inner drive and energy that she was no longer able to express. I tried to picture in my mind what she would have been like now had she received natural hormone therapy before all these permanent changes occurred. Her devastating situation deeply influenced my determination to write this book.

Another account that fueled my interest in sharing this needed information came from a dear friend in her seventies. She was a former surgical nurse and, needless to say, had always been health conscious, trying to follow her doctors' orders to the letter. Well, my friend had just heard from her orthopedic surgeon the shocking news that she had quite a lot of bone loss. He proposed that she be thinking about a hip replacement. She recalls driving home with tears of disbelief at this news thinking: "How could this be, when I've always taken preventive cautions so as to avoid this dreaded diagnosis of osteoporosis?"

She finally telephoned me and said, "I don't understand why I'm having bone loss. I've been taking my estrogen for over twenty-five years, thinking that this was good for my bones." How do you tell a nurse that she has been on the wrong hormone all these years? And that estrogen's role is only to slow down the bone breakdown, while progesterone is the hormone that plays the major role in building bone mineral density? As she read a draft of this chapter and learned of the importance of progesterone as it works on osteoblast cells to increase bone mass, she related her ambivalent feelings. On the one hand, she felt robbed of her health because she had used the wrong hormone for so long; but on the other, she was relieved to realize her problems could be reversed.

If we recognize the signs of hormone imbalance and understand how and why hormone replacement therapy can correct such an imbalance, we'll be spared a lot of grief in years to come. Who wouldn't get tired of going from one doctor to another for headaches, fatigue, heart palpitations, hot flashes, heavy bleeding, irregular periods, spotting, anxiety, moodiness, night sweats, depression, or just feeling rotten? Many doctors still don't recognize that many of these problems are related to menopause or other stressful times—such as after childbirth or hysterectomy.

Just like the woman in the story above, many of us simply endure these symptoms and learn to live with them by keeping busy, trying to eat right, and so on—completely unaware of the harmful effects of progressive progesterone deficiency. Medicine is triumphant in emergencies and trauma, but it often falters in the management of preventive health care. The very drugs your doctor prescribes may mask an underlying hormone deficiency. Sooner or later, when the symptoms get bad enough, you may have to supplement Mother Nature's gifts with what she is no longer adequately providing. Why not use a natural progesterone product?

HOW MANY TRAGEDIES WILL STRIKE BEFORE WE TAKE A STAND?

No wonder osteoporosis is so prevalent. Have we indeed been given the wrong hormone—a hormone that medical doctors have been prescribing for over five decades? Since 1942, when estrogen was approved by the FDA for production by Wyeth-Ayerst Laboratory and introduced to the market, it has been demonstrated to be of little worth in the prevention of osteoporosis. It seems that women have been under a theoretical HRT system that has been extensively prescribed without necessary checks and balances. No wonder osteoporosis is called "the silent killer." There comes a time when we may become too old to feel comfortable voicing complaints, and our bone degeneration, fractures, and joint problems continue as a generational plague.

FLUORIDE: HARD TO AVOID

Try as we may to create all the right conditions to avoid osteoporosis—the right kind of hormones, the right kind of calcium, and the nutrients needed for its assimilation—we might also want to be more aware of a chemical that is added to our water and of its effect on bone mass. The threat posed by fluoride may very well obliterate all our good intentions.

The *Holistic Dental Digest* tells us that in addition to devitalizing the tooth enamel and possibly leading to periodontal disease, fluoride actually causes the bones to become more brittle and weak—even though they *appear* denser on X-ray films. One study reports that "the risk of fractures for women in high-fluoride communities [is] more than double."[94]

Another study reviewed in the *Journal of the American Medical Association* found that fluoridation increases the rate of hip fracture by about 30 percent in women and 40 percent in men.[95] Despite false claims by some public health officials to the contrary, there are no legitimate studies to indicate any hip fracture protection from fluoride. Fluoride is toxic to bones in any amount, including the level found in fluoridated water.[96]

It is also harmful to the thyroid and a known cause of cancer and other diseases. Japanese researchers reported in 1982 that "sodium fluoride, which is being used to prevent dental caries [cavities], produces chromosomal aberrations and irregular synthesis of DNA." The latest studies in the United States confirm the frightening truth that "malignant transformation of cells is induced by sodium fluoride."[97]

Over three decades ago we were warned of these dangers in *The American Fluoridation Experiment* by Frederick B. Exner, M.D., an X-ray diagnostician, biologist, chemist, physicist, and pathologist.[98] And the American Medical Association published data in the *Archives of Environmental Health* in February 1961, showing that fluorides have been found in diseased tissue from tumors, the aorta, and cataracts.[99]

Yet, from that day to this, says *Cancer Forum*, the U.S. Public Health Service has encouraged the Environmental Protection Agency to keep assuring American citizens that fluoridation is effective and safe, in spite of the fact that there is "not a shred of scientific evidence to support that claim."[100] It would be in our best interest to mount a national letter-writing campaign to convince Congress and the EPA to remove this hazard from our public water supply.

MISCONCEPTIONS ABOUT CALCIUM AND DIET

When we are trying to sort out the facts on how to maintain an adequate calcium level, let's acquire our information from unbiased sources rather than from those associated with the dairy, cattle, and poultry industries. A dramatic example appears in *Health Science*, where Dr. Joel Fuhrman states: "If your calcium intake is very high but you constantly excrete more calcium than you absorb, you are in a negative balance and osteoporosis will result in time. On the other hand, if your calcium intake is relatively low but your body is efficient at absorbing it, you are in a positive balance and your skeleton will not be stripped of its calcium stores."[101] Drs. Cedric and Frank Garland *(The Calcium Connection)* and Nancy Appleton, Ph.D. *(Lick the Sugar Habit)* point out that excessive dietary sugar intake increases calcium excretion in the urine, upsetting the body's mineral balance.[102]

From an academic viewpoint, when we refer to calcium and other minerals in our chemistry classes, the words "organic" and "inorganic" have a strictly applied definition. However, writers on nutrition use the term *organic* more loosely to mean *alive* or *bioavailable*. It is in this sense that numerous studies have been made regarding organic and inorganic minerals and their use to, or abuse of, the body.

Dr. M. T. Morter, Jr., cautions against depending on unreliable types of calcium. He is convinced that the body cannot dissolve the strong "ionic" bonds of "inorganic" calcium, such as dolomite or oyster shell. Nor, in spite of dairy industry advertising, can it utilize the calcium found in cow's milk; because unless the milk is consumed raw, its calcium has been altered by the pasteurization process into a hard, unusable form, which will be deposited in the wrong areas of the body. Dr. Morter points out that women who consume large amounts of dairy products are actually among the high-risk groups for osteoporosis.[103]

Along this same vein, Norman W. Walker, D.Sc., in his book *Fresh Vegetable and Fruit Juices*, gives us some excellent thoughts to live by concerning oxalic acid and its relation to calcium assimilation:

> When food is raw, whether whole or in the form of juice, every atom in such food is vital [i.e.,] ORGANIC and is replete with enzymes. Therefore, the oxalic acid in our raw vegetables and their juices is organic, and as such is . . . essential for the physiological functions of the body. . . . The oxalic acid in cooked and processed foods, however, is definitely dead, or INORGANIC, and as such is both pernicious and destructive. Oxalic acid readily combines with calcium. If these are both organic, the result is a beneficial constructive combination, as the former helps the digestive assimilation of the latter, at the same time stimulating the peristaltic functions in the body.[104]

Please note that this advice is contrary to what you may have read elsewhere, but it really makes sense—and its author has much experience as well as credibility. Dr. Walker, after all, lived to be one hundred and nine years of age by practicing what he preached; his wife died in her nineties. He is the only person I've known to explain the distinction between *raw* and *cooked* foods containing oxalic acid:

When the oxalic acid has become INORGANIC by cooking or processing the foods that contain it, then this acid forms an interlocking compound with the calcium even combining with the calcium in other foods eaten during the same meal, destroying the nourishing value of both. This results in such a serious deficiency of calcium that it has been known to cause decomposition of the bones. This is the reason I never eat cooked or canned spinach.[105]

Dr. Walker tells us that the most plentiful quantities of organic oxalic acid are found in fresh raw spinach, kale, collards, mustard greens, turnips, Swiss chard, and beet greens.[106] Other sources include almonds, asparagus, and parsley.

Many of these same foods are high in calcium as well, and indeed, the first source we should look to for our calcium requirements should be our daily diet. But as we've seen from Dr. Morter's views earlier, the customary idea of milk as one of the most important sources of calcium is coming into much debate these days. In fact, you will assimilate more calcium from ingesting kale than from drinking milk, according to Frederik Khachik, a researcher at the U.S. Department of Agriculture.[107]

It would seem that the very foods we are told to consume for calcium are those that often cause allergies and other serious problems. By way of illustration, cardiologist Kurt Oster conducted extensive research into the xanthine oxidase in homogenized cow's milk. This substance was shown to damage arteries and promote atherosclerosis. He found no such correlation associated with the intake of butter or cheese, presumably because they contain little or no biologically active xanthine oxidase. Oster makes a reasonable case that one of the causes of atherosclerosis is the consumption of homogenized milk.[108]

There are valid concerns, too, about skim milk. Butterfat actually provides an excellent source of vitamins A and D, is anticarcinogenic, and allows the minerals and other nutrients in milk to be absorbed and utilized. One researcher observes, "The plague of osteoporosis in milk-drinking western nations may be due to the fact that most people choose skim milk over whole, thinking it is good for them."[109]

Synthetic vitamin D, added to replace the natural, is toxic to the liver. Nonfat dried milk, which is added to 1 percent and 2 percent milk and to nonfat commercial yogurt, contains heart-damaging rancid cholesterol and high levels of nitrites and galactose (milk sugar linked with development of cataracts, glaucoma, and ovarian cancer). Thus, for several reasons, whole milk products may be preferable.[110]

A further commentary on milk comes from *How to Get Well*, a useful health book that provides practical information on nutrition and drugless treatment. Dr. Paavo Airola writes:

> Today's pasteurized supermarket-sold milk is loaded with toxic and dangerous drugs, chemicals and residues of pesticides, herbicides and detergents—such milk is not suitable for human consumption. If you are fortunate enough to get *real* milk, fresh, raw, "farmer" milk from healthy cows fed organic food, then you can add milk to your diet.[111]

To his list we might add the hormones that are often present in commercial milk products. Dr. Airola says that the preferred way of consuming milk is as acidophilus milk, yogurt, or other soured (predigested) forms, because they "help to maintain a healthy intestinal flora and prevent intestinal putrefaction and constipation."[112] Look for plain, unsweetened yogurt (preferably organically produced) with active lactobacillus cultures or for cultured buttermilk. In some places unhomogenized commercial yogurt is available; homemade whole milk yogurt would be even better.[113]

However, if you do decide to use a calcium supplement (particularly after menopause), many authorities recommend one such as calcium citrate[114] over the more common calcium carbonate. The latter can be hard to assimilate, especially in older persons and those deficient in hydrochloric acid. But remember also the importance of magnesium and the trace nutrients mentioned earlier in the chapter, which facilitate calcium uptake by the bone.

Assimilable calcium and magnesium can now also be found in a micronized spray form that is absorbed by the buccal mucosa (inner lining of the cheeks) in fifteen to twenty seconds and directly enters the bloodstream, thus eliminating digestive system involvement. This may provide us with an entirely new means of meeting our daily requirement for these critical minerals. The trace minerals that also play an important role in optimal bone nutrition are now available in various liquid or spray forms as well.

BUILDING BONE WITH B VITAMINS

And what else do we need to know about the health of our bones? Dr. Alan Gaby stresses the importance of providing structural proteins to bone tissue. He discusses the influence of vitamin B_6 in this area, pointing out that "this vitamin is a cofactor for the enzyme lysyl oxidase, which crosslinks proteins and connective tissue. Adequate vitamin B_6 is therefore required to provide tensile strength and structure to collagen . . . in bone tissue." He also mentions a study by Nelson, Lyons, and Evans, which "suggests that vitamin B$_6$ enhances the production or the effectiveness of progesterone. To the extent that progesterone is important for bone health, adequate intake of vitamin B$_6$ is also essential."[115]

THE IMPORTANCE OF WEIGHT-BEARING EXERCISE

The Center for Women's Health at Columbia-Presbyterian Medical Center surveyed women who exercised three or more times a week and found that they reported fewer osteoporotic symptoms than those who exercise twice a week or less.[116] But what kind of exercise would help us prevent osteoporosis? The importance of an exercise program is discussed in Dr. Gaby's book *Preventing and Reversing Osteoporosis.* He states that weight-bearing exercise, which forces the body to work against gravity, helps in building bone density. For older women, swimming may be most appropriate, as there is less risk of injury.[117]

Dr. Gaby observes that urinary calcium excretion was found to be quite extensive in astronauts following their time in space. Furthermore, patients restricted to bed for back pain demonstrated rapid bone loss: "bone mineral content of the lumbar spine decreased at an astounding 0.9% per week." Dr. Gaby emphasizes, therefore, that "physical activity plays a crucial role in maintaining bone mass."[118] (For further information, see pages 56 and 93.)

The role of exercise in preventing and reversing bone depletion cannot be overstated, nor can exercise alone make all the difference. As we have mentioned again and again, it is the combination of factors that has the greatest impact on our total health and specifically on our bone mineral density. In planning your exercise program, remember that it, like dietary changes, needs to be incorporated into your daily routine—for the rest of your life. Just as crash diets yield only temporary weight loss, sporadic exercise is not going to have any significant impact on your bone density.

Your body will give you clues as to the right kind and amount of exercise for you. Dr. Ray Peat points out in *Nutrition for Women* that women athletes are sometimes chronically deficient in progesterone (and may miss periods), probably because the stress of hard exercise causes the conversion of progesterone to cortisone.[119] He also observes that excessive stress and so-called aerobic exercise that leaves one breathless are common causes of hypothyroidism[120] and may contribute to premature aging. Thus, we must find that happy medium.

Here are a few guidelines to keep in mind when planning an exercise program:

- Start slowly and steadily; the no-pain/no-gain motto does not apply. If a twenty-minute walk causes back pain, for example, don't stop altogether; just cut the total time in half for the next week or two. Find your comfort level and build gradually from there.
- Aim for a thirty- to forty-five-minute work-out period at least three times per week.
- Include a combination of aerobic exercise (walking or jogging), strength training with light to moderate weights, and a stretching or yoga routine for flexibility. This provides variety and helps you avoid boredom.
- Check out books, tapes, and videos at your local library, bookstore, or video store; this way you can expose yourself to a wide variety of trainers and instructors.
- And for chronic exercise avoiders, here's a no-excuse, no-reason-not-to, simple-as-can-be weight-bearing exercise—the bone bounce: simply rise up on your toes and let yourself drop down on your heels, making sure you feel a vibration on your hips. You can vary the intensity as needed by varying how high you rise up off the floor. Do this two hundred to three hundred times or for two to three minutes per day.
- Finally, you're never too old or too young to start an exercise program; just remember to begin slowly, build gradually, and check with your health-care providers if you have any medical, neurological, or musculoskeletal complications. Even seventy-year-olds in wheelchairs have improved their flexibility and overall fitness with light weightlifting.
- Most of all, be creative and make it fun.

WHERE DO WE GO FROM HERE?

It is not difficult to be overwhelmed by the health decisions that confront us. Modern-day medicines include never-ending lists of names for "new and improved" synthetic drugs and hormones. (See appendix B for a sampling of the many synthetic hormones commonly prescribed today.) Because the terms for the complex chemicals we are consuming are often incomprehensible, we too often rely completely on our medical doctors for treatment decisions.

What a contrast, and what an awakening, to discover information that is understandable! It seems almost too simple to be true—but then, isn't truth often found in the simpler forms of nature?

What we have learned so far may require some assertive action. With alternatives to artificial hormones, we now have an opportunity to become less dependent on drugs that lead us away from a sound body, mind, and soul. Prepared, we can responsibly choose a path less traveled—one that may lead to a better quality of life.

In appendix F, you will find a form letter you can give to your doctor explaining your concerns about synthetic hormones. He or she should know how you feel about toxic drugs and the unwanted side effects of the synthetic hormones that cause so much distress. It is imperative that your health provider understand that your body *does require*, and *can assimilate*, natural progesterone.

It is important to note that the cream used in Dr. Lee's studies specifically contained USP progesterone, which is referred to as "natural" progesterone in most of the current literature. We are often asked if there is any difference between this and the many wild yam creams available today. The answer is that, to date, no large-scale clinical studies have been done to validate the use of wild yam extract without USP modification.

With respect to the treatment and prevention of osteoporosis, so far only USP progesterone has been clinically tested and evaluated. However, for the treatment of many of the other debilitating and draining symptoms of PMS and menopause (cramping, irritability, hot flashes, fibroids, etc.) patient after patient reports relief using a product containing the sterol from either the wild yam or soybean without conversion to the USP progesterone. Contrary to much of the skepticism regarding these creams, many appear to be bioactive and able to promote hormonal balance. (See appendix A for more information). Some women have found that alternating the two types (or using a cream that combines both) gives the best results. (See appendix G.) Others alternate the cream with sublingual drops or tablets.

⌇

Dr. Judi Gerstung is interested in assessing long-term response to various treatments for increasing bone mineral density. If you would like to assist her with this project, please send a brief description of your health-care program, a copy of your initial bone mineral density test, and results of any follow-up studies: Dr. Judi Gerstung, P.O. Box 7149, Marietta, GA 30065-0149 (e-mail: judigerstung,dc@hotmail.com).

CHAPTER 5

THE RISK OF CANCER

The evidence is strong that unopposed estradiol and estrone
are carcinogenic for breasts. . . .
The only known cause of endometrial cancer is unopposed estrogen.
John R. Lee, M.D.

W e've been waging the war on cancer for many decades, and what do we have to show for it? Despite the vast sums of money that are spent on developing new drugs and new technologies, casualties continue to mount. Dr. Allen Astrow, a cancer specialist at St. Vincent's Hospital and Medical Center in New York, tells us that even with all the advances in our knowledge of the molecular mechanics of tumor formation and development, "mortality rates from cancer in the U.S. are rising."[2, 3]

Once we begin to understand the reason cancer exists, we will see why the war against cancer was lost years ago and we need to get off this battlefield. Dr. John Lee alerts us to the research of Schippers et al., which shows that "cancer cells, far from being foreign invaders, are an intimate part of ourselves, essentially normal cells in which proportionately small changes in genes have led to changes in their behavior. The treatment strategy should be to reestablish intercellular communications." It's called *rebalancing*.

I was able to understand the process of rebalancing the body a lot better after reading Dr. Richard Passwater's research on human cancers and our environmental carcinogens. Thousands of new chemicals are added to the environment every year.[4] A study by the Department of Biological Sciences at the University of Idaho showed that many environmental pollutants contain estrogenic compounds that are suspected of increasing the occurrence of endocrine interference and reproductive failure in humans and other species. According to Dr. Raymond Peat, in a personal communication, "environmental carcinogens—phenolics, aromatic-hydrocarbons, chlorinated-hydrocarbons, PCBs, dioxins, soot, even X-rays—are in general estrogenic." Although people are only beginning to be aware of this deadly connection, especially to breast cancer, it has been published widely in scientific journals and the popular press.

HORMONES, GENETIC MUTATIONS, AND CANCER

The topic is controversial and in many respects frightening. Yes, it seems that a large number of the chemicals we encounter in our daily lives, such as pesticides, solvents, industrial contaminants, and components of detergents and plastics, actually mimic estrogen in living organisms. Because these harmful estrogenic substances latch onto or sometimes enter our living cells, which contain receptors, they can either give the cells a greater than normal hormonal "charge" or block activity of the body's own estrogen by jamming the receptors.

Why is this so serious? Recent studies cite the effects of such environmental contamination: a dramatically rising breast cancer rate in the U.S., a possible decline in sperm quality plus low testosterone levels in men, abnormal penises and reduced motor and language skills among boys whose mothers were exposed to estrogen-like toxins, and other ills. Not only are humans threatened, but some animal populations (marine birds, panthers) are perhaps very close to extinction. Because of the estrogenic pollutants, sea gulls and alligators have been found with abnormal reproductive organs and a low reproduction rate, with nesting impulses observed among males and aggressive behavior in females.[5,] *

Instead of degrading like other poisons, substances such as DDT, PCBs, and dioxin accumulate over the years. Since some of these fat-soluble chemicals are sprayed on plants and eaten by animals, they especially affect those at the top of the food chain: humans. What does that bode for the next generation?

Ovarian, breast, prostate, and testicular cancer—all hormone-driven—have increased sharply in recent years. Now the Environmental Protection Agency and several other organizations are looking into the impact of estrogen-mimicking chemicals on both reproduction and cancer, in an attempt to either confirm or deny this compelling theory.

---

* To read more about this, see Thoe Colborn's "Developmental Effects of Endocrine Disrupting Chemicals in Wildlife and Humans" in *The Journal of the National Institute of Environmental Health Sciences*, Vol. 101, No. 5, October 1993.

It seems quite convincing when it strikes close to home. I myself know of a bright young college student, ironically an environmental health science major, who at the age of twenty-three has endured multiple surgical procedures and faces uncertainty about whether she will ever be able to bear children—all as a result of exposure to the pesticide Safrotin, which was sprayed, improperly, inside her dormitory room.

Five months after an unpleasant and violent initial reaction, the hormonal effects of this neurotoxic chemical began their toll: midmonth bleeding, excruciating cramps, diarrhea, and urinary pain. Before finally arriving at the diagnosis of severe endometriosis, a series of doctors put this young woman through a terrible ordeal of testing, exploratory surgery, and ineffective treatment. Then they gave her undesirable options such as synthetic birth control pills to try to normalize her cycles, more surgery to determine the extent of damage to her ova, and treatment with a drug that would induce menopause.

In view of the potentially tragic health consequences, Dr. John Lee feels we should be alarmed about our exposure to these foreign xenoestrogens (or "xenobiotics") derived from petrochemicals. Indeed, he says there is a corresponding epidemic of progesterone deficiency among women in the industrialized nations (North America and Western Europe), influenced in part by environmental contamination and our processed food supply, whereby we do not get the progesterone in our diets that people in the less developed countries obtain from raw, natural foods.[6]

Referring to the fascinating work of Dr. Peter Ellison for the World Health Organization, Dr. Lee says:

> The estrogen levels in America today tend to be quite a bit higher than in other countries. With menopause the fall of estrogen is greater in the U.S. and other industrialized countries than in the more agrarian "Third World" countries. Progesterone levels are remarkably stable in agrarian countries whereas, here in the U.S., women's progesterone falls to levels quite close to zero, even lower than that found in men.[7]

Estrogen Overkill

In short, all women are in danger of estrogen excess. Consider the sources: (1) Estrogen is secreted in substantial amounts by the ovaries and adrenal glands in the form of androstenedione (an estrogen precursor).[8] (2) Medical doctors prescribe estrogens for numerous clinical conditions at many different stages and ages in a woman's life cycle—for menstrual problems, infertility, PMS, birth control, and postmenopausal symptoms. A big concern is that we are thereby adding more estrogen to an already estrogenic environment, setting our bodies up for a multitude of disorders and diseases. (3) Not only are natural estrogens present in many plant foods (phytoestrogens), but if we eat meat and dairy products that are not organically produced, we may also be ingesting foods daily that contain synthetic estrogen (DES)[9] or the growth hormones that are injected into cattle and poultry. (4) And finally, we live in a chemically saturated environment, which is a source of additional undesirable estrogen.

With the cards stacked against us like this, most of us will undoubtedly experience what Dr. Lita Lee has referred to as "estrogen overload."[10] The host of contributing factors makes it fairly easy to understand how estrogen can permeate and eventually gain the upper hand in the body, doing significant damage without its essential partner, progesterone, to balance and oppose it. Even herbal estrogen, when taken alone, can contribute to this type of hormonal imbalance.

Our human cells take an amazing amount of abuse from artificial preservatives, colorings, and flavorings and the other chemical additives in our foods. Just think about the animal feed additives, antibiotics, pesticides, drugs, and hormones that are given to animals with little or no federal regulation.[11]

SPECIFIC ESTROGENS AND CANCER

Many experts have observed a direct link between breast and uterine cancer and high levels of particular estrogens in the body—specifically the estrogens known as *estradiol* and *estrone*. The other common estrogen, *estriol*, is now thought by some (although the matter is controversial) to be the "good guy." This was brought home to me by an eye-opening piece of information I came across during my search for natural hormones. The statement in Dr. Julian Whitaker's newsletter *Health & Healing* still sticks in my mind. At first, like a kid who's just come home from school with a new and exciting bit of knowledge, I had to quote it to all my friends:

> There are three forms of estrogen: estrone, estradiol, and estriol. It was shown almost 10 years ago that estrone and estradiol were the primary cancer culprits, while estriol was actually associated with a reduced cancer incidence. Premarin is composed primarily of estrone and estradiol.[12]

Estriol is most dominant in a woman during pregnancy when very large quantities are produced in the body, along with very low levels of the other two common estrogens. Dr. Lee reports that a study of women with breast cancer found they had 30 to 60 percent less estriol in their urine than did a control group of women without cancer. Although this phenomenon could be related to the reduced liver function of most cancer patients and not to any cause-and-effect relationship, it has been surmised that supplements of estriol might block the carcinogenic effect of excess estradiol and estrone and act, in effect, as a cancer-preventive (or anticancer) agent.[13] Unfortunately, the lack of agreement among scientists on this theory stems from the scarcity of research data.

The situation with endometrial cancer is similar. A study reported by the *Journal of the American Medical Association* showed estrone to be the principal component in the induction of endometrial carcinoma.[14] And a respected researcher tells us that the only known cause of such cancer is the presence in the body of excessive quantities of certain estrogens (in particular estrone and estradiol), accompanied by low levels of estriol and unopposed by the counteraction of adequate progesterone.[15, 16]

It's a sad fact that the approved and medically preferred estrogens prescribed orally to women are combinations of estrone and estradiol, or estradiol by itself;[17] and these are the very forms traditionally used as supplements to combat osteoporosis.[18] The grim consequences of these hormones also befall menopausal women who are on short-term supplementation during the usual five-year period of menopause. Such estrogen supplements have been shown to increase the risk of hormonal-driven cancers sixfold—with longer-term use multiplying the risk by as much as fifteen times that of non-users.

Because of the known risks of the synthetic products, many doctors open to natural alternatives now believe that supplemental estrogen is not necessary as long as a woman is receiving natural progesterone. Since progesterone is a substance from which estrogen is formed—that is, a precursor of it and other hormones (cortisone, testosterone, and aldosterone)[19]—under normal conditions it will stimulate production of one's own estrogen.

Why, then, are the medical doctors continuing to give us synthetic estrogens? Why do so few physicians prescribe the wiser choice of natural progesterone instead of synthetic progestins? And how do we, the recipients of such drug treatments, make sense of this baffling approach?

I can say from my own research and experience that for those who have been through menopause or have had their uterus or ovaries removed, the need for a safe approach to estrogen supplementation seems glaringly obvious. The case seems pretty clear: breast and endometrial cancer are caused to a great extent by excessive, unopposed levels of two particular estrogens, estradiol and estrone. Is estriol in combination with natural progesterone, then, the answer? Certainly the addition of progesterone would be an important facet of cancer prevention because it has been shown to stop cells from multiplying.[20]

IS THERE A SAFE ESTROGEN?

An article that was buried in a 1978 issue of the *Journal of the American Medical Association* states that "Estriol may not only be noncarcinogenic but indeed anticarcinogenic. [A] high endogenous estriol level protects against the tumor-producing effects of estrone and estradiol." It further goes on to document a "notable inhibition of mammary carcinogens with estriol therapy compared with therapy using estrone and estradiol" (currently being used by most doctors under the names Estraderm, Estrace, and others).[21]

The estriol trials as described in this article resulted in 37 percent of the postmenopausal patients with breast carcinoma and metastases demonstrating remission or arrest of metastatic lesions—with a dose of "2.5 to 5 mg and occasionally 15 mg of estriol, equivalent to a little more than 0.65 and 1.25 mg of conjugated estrogens."[22]

When I read in this same article that "orally administered estriol is safer than estrone or estradiol" and that diethylstilbestrol, the hormone mainly prescribed for advanced cancer, "is not a steroid but is a chemical complex that acts like estrogen and is as carcinogenic as estrone,"[23] I wondered why most gynecologists or other medical specialists never mention these things. Nevertheless, Dr. John Lee's work on NHRT does give women this information to add to our treasury of knowledge:

> The addition of progesterone enhances the receptors of estrogen, and thus [a woman's] "need" for estrogen may not exist. If neither vaginal dryness nor hot flushes are present after three months of progesterone therapy, it is unlikely that estrogen supplements are needed. . . . Once progesterone levels are raised, estrogen receptors in these areas become more sensitive, and hot flushes [flashes] usually subside. The validity of the mechanism can be tested by measuring FSH [follicle-stimulating hormone] and LH [luteinizing hormone] levels before and after adequate progesterone supplementation.[24]

Finding all this information was enlightening. However, first I had to experiment on myself. A pragmatist by nature, I then saw my goal become clear: to see whether these two hormones, progesterone and estriol, would put an end to my own problems. After I had read books, articles, abstracts, reports, and studies to find what was available on the subject, and—after contacting Emory University, pharmacies, medical centers, and reference libraries for pertinent documentation—I started on NHRT.

During and after the six months of my experiment, I felt better than ever before, so my second goal became obvious. I wanted to reach as many other women as possible to encourage them to take action on behalf of their own health and well-being, to consider any as-yet unknown long-term effects, and to evaluate and monitor their bodies' reponses to the treatment of their choice. In this way, they could perhaps gain as much as I, relieving not only their fears and frustrations but their mental and physical disorders as well.

Synthetic Hormones Create Imbalance

Wallace Simons, R.Ph., of Women's International Pharmacy told me what happens when progest*ins* do not function appropriately at the sites of progest*erone* receptors in target cells throughout the body. It is as if these synthetic components are blocking the progesterone sites. As Dr. John Lee describes it, the body perceives that since the progesterone sites are all occupied, its own natural progesterone (assuming the woman is still ovulating) cannot be utilized, and therefore production should be cut back.

Without the adequate opposition of progesterone, estrogen then becomes dominant and runs rampant. This in turn causes the body's feedback mechanism to recognize that something is still needed in the body's many progesterone receptors (as discussed in chapter 4). A message is sent to the ovaries to increase their own progesterone output to fill the receptors. But because these sites are unavailable, this additional progesterone cannot be used properly. The woman plunges into bodily chaos, which affects her emotions, memory, sex drive, body temperature, and other physiological functions that mirror hormonal decline.

Excessive environmental estrogens and the prescription of additional synthetic estrogens and progestins may play an underlying role in the stimulation and rapid multiplication of cells (often in breast or cervical tissues) and lead to hormone-fed cancers. Dr. Ray Peat explains another reason this happens: "The thymus gland is the main regulator of the immune system. Estrogen causes it to shrink, while progesterone protects it." Furthermore, he says that estrogen excess actually obstructs oxygenation of the blood.[25] This is significant in view of Dr. Otto Warburg's finding that "normal cells use oxygen-based reactions as their source of energy, but cancer cells can form from cells not receiving adequate oxygen."[26] The body becomes overwhelmed and run down by the resulting imbalance, and the immune and circulatory systems are weakened.

In an important study reported by Johns Hopkins University in the *American Journal of Epidemiology*, women were first measured for their estrogen and progesterone levels and then placed in two separate groups to monitor their susceptibility to cancers. The findings proved quite interesting:

> A test was run for 40 years. They found when the "low progesterone" group was compared to the "normal progesterone" group, the women in the low progesterone group had [approximately] $5^1/_2$ times the risk of breast cancer.[27]

A further conclusion was that the "low progesterone" participants encountered a tenfold increase in deaths from *all* types of cancer, compared with the "normal progesterone" participants. This should be headline news! However, according to Dr. John Lee, these important test results were not distributed or further publicized. Why? Can you imagine the thousands of women who might have avoided the dreaded disease of cancer, and even the difficulties of menopause, through natural prevention? Instead, women were given little hope—or choice—by their doctors except the radical methods of radiation, chemotherapy, and/or surgery.

There is a much simpler and saner way: balance the hormones so the body's own defense system kicks in, stopping disease in its tracks before it gets started. Indeed, progesterone alone seems to be the hormone of choice for most conditions. If, however, estrogen supplementation seems unavoidable, it appears from some reports that estriol, in harmony with natural progesterone, may help to neutralize any excess or correct a deficiency of some of our other hormones.[28–30] Is this natural hormone replacement team one of the world's best-kept health-care secrets? Perhaps, but Dr. Ray Peat cautions that there is evidence that estriol and other phytoestrogens can add dangerously to the body's total estrogen load.[31]

BREAST CANCER

Is it any great mystery that more and more women are succumbing to cancer at such a rapid rate? Just look at the statistics, which speak for themselves: Breast cancer is the leading cause of cancer death in women in the United States. According to *Morbidity and Mortality*, the report published by the Centers for Disease Control, the incidence of breast cancer increased 52 percent between 1950 and 1990.[32] Ninety thousand new cases are diagnosed annually, and 37,000 women per year will die of it.[33]

If natural progesterone therapy were given as much commercial time as is mammography for preventing breast cancer, women would certainly have a greater opportunity to learn the true meaning of prevention of disease. And when it comes to progesterone, it seems reasonable that what's good for the development and health of the embryo (progesterone as the dominant hormone) would also be a good equalizer for women going through the many changes and stresses in their lives. To understand why this is, take a look at Dr. Lee's response to a question asked of him during an interview by Ruth Sackman, president of F.A.C.T. (Foundation for Advancement in Cancer Therapy):

> We know that a woman is protected from having breast cancer if she has multiple pregnancies. In multiple pregnancies you have long periods of time where progesterone is the dominant hormone. In breastfeeding the ovaries do not start raising estrogen. So if a woman combines pregnancies with some time of breastfeeding, her breasts will be much protected against the estrogen effects.[34]

Fortunately, Dr. Lee's work with cancer patients who at the same time were suffering from osteoporosis has laid the foundation for a transformation in how today's women will approach menopause.

The lesser-known hormones have been quietly written about for years. But women at large, who don't have ready access to medical journals, have not been made aware of the implications. Fortunately, however, Dr. Lee discovered the importance of natural hormones as far back as 1981. For more than fourteen years he observed the results of progesterone therapy in his patients and saw absolutely no return of their cancers. By making known to us his lifesaving findings, Dr. Lee has given women worldwide the benefit of his efforts.

PROGESTERONE RECEPTORS AND BREAST CANCER

This may be the time to bring up a very important point. Many readers who have had breast cancer, and who have educated themselves and now want to use natural progesterone with their doctors' blessing, are going to run into a very common brick wall. If a cancer patient has undergone breast biopsy, more than likely she has also had the standard estrogen and progesterone receptor assay. This test is used to measure the quantity of hormone receptors on the tumor cells and then, theoretically, to serve as a guide to the advisability of hormone treatment of any kind. If the results show that the number of progesterone receptors is high, most doctors will not prescribe progesterone. They're unsure of how to interpret the test results.

After getting this same response from several doctors, even the more open-minded ones, one patient took matters into her own hands and wrote to the company in California that had run her tests, asking for the medical abstracts on which the evaluation was based. The only articles available turned out to be over a decade old and seemed to conclude that more study was definitely needed.[35-37] The laboratory did, however, admit that the assays are performed using *synthetic* progestins—leading one to wonder just how much value they hold! The woman's gynecologist was very surprised to learn this.

Obviously she is not alone, because Dr. John Lee is asked about this dilemma all the time. He wrote to her, "Like many other things in life, we are forced to make decisions despite incomplete knowledge. . . . The available knowledge concerning the significance of hormone receptor status is still a bit murky, [and] the testing procedure itself involves some uncertainties." First of all, he pointed out, "One cannot assume that tumor cells are all alike."

Dr. Lee, in his explanation to the frustrated woman, said that progesterone receptors (PgRs) "do not develop in breast cells (or tumor cells) unless estrogen is present in good quantities. The very presence of a high number of ERs [estrogen receptors] and PgRs is a good sign, in the sense that the tumor cells are more like fully differentiated normal cells."

Estrogen acts to increase cell proliferation. Progesterone, on the other hand, acts to induce cell maturation and differentiation. Therefore, said Dr. Lee, "the presence of PgRs allows the hypothesis that the addition of progesterone makes the tumor cells more mature and more differentiated, creating a lower level of malignancy among them. The authors of the papers [on the ER/PgR assay] fail to consider the different actions of the hormones involved."

So his advice to the doctor of the breast cancer surgery patient is this: "From all the evidence I've seen, even if . . . the receptor sites are there for progesterone, the patient is a perfect candidate to use progesterone. . . . That's the only way the progesterone could ever work. Whereas if it's estrogen-site positive, then she should not have estrogen because it causes the cell to multiply. What does progesterone do? It causes [the cell] to stop multiplying."[38]

According to Dr. Lee, "My long experience in using natural progesterone in patients with a past history of breast cancer treated by surgery alone, and the finding that none of them have shown late metastases or recurrences, leads me to believe that natural progesterone poses no increased risk to such patients . . . [and is] quite probably a benefit." He concluded, "If the PgR test is negative, the cell will be unaffected by progesterone. If [it] shows a good quantity of PgRs, supplemental progesterone may make the tumor cells less malignant."

And now, let's take a look at some risky practices associated with either the cause or the treatment of hormone-related cancers, and the forces that are at play. On the one hand we have the environmental, synthetic estrogens and the bad estrogens that could dominate the body (in Dr. Lee's conviction these are the estrogens that are "the sole cause of fibrocystic breast disease [and] the only known cause of cancer of the uterus"[39]).But on the other hand, to bring about balance, we have plant-derived progesterone and what may or may not turn out to be "good" estrogens—such as estriol and other forms that come from plants. What follows, however, is what we're up against if *not* given the choice of natural hormone replacement therapy.

Avoidable Hysterectomies and Their Aftermath

Hormone deficiency is a fact of life that we must address in order to understand the problems that may begin at childbearing age. The slow decline of progesterone, such an essential hormone, is overwhelming enough; but when there is a sudden halt, it can be devastating. A complete hysterectomy, in which the ovaries are removed, can create numerous problems, such as adrenal and pancreatic disorders (e.g., diabetes) and other side effects, including vaginal dryness and sudden hot flashes.[40]

Estrogen is traditionally prescribed at this time, but as Dr. Lita Lee comments, "This seems such a dichotomy to me, since estrogen excess and/or progesterone depletion is one of the primary causes of problems leading to hysterectomy (excessive bleeding, fibroids, endometriosis and cancer). The more research I read, the less inclined I am to think that women need supplemental estrogen even after menopause."[41]

If the doctor prescribes synthetic estrogen, it won't be long before one or more of the following conditions inevitably arises: abdominal cramps, interruption of periods, bloating, breast tenderness and enlargement, cystitis, high blood pressure, endometriosis, gallbladder disorders, hair loss, elevated blood fats, jaundice, depression, nausea and vomiting, sustained vaginal bleeding, decreased carbohydrate tolerance, skin outbreaks, blood clots in the legs, unwanted weight loss or gain, or vaginal yeast infections.[42]

Any woman who develops these problems has really been mistreated fourfold by the doctor prescribing the estrogen: (1) It is unopposed by natural progesterone, (2) synthetic estrogen can in the long run cause problems worse than the original complaints, (3) Provera and other imitations of progesterone, if prescribed, exaggerate the symptoms even further still, and (4) all of the above conditions not only accelerate the aging process but open up an environment favorable to cancer.

When such symptoms appear, some action may be necessary before a more severe stage of development, such as a tumor, occurs. We need to watch and listen for "signals," be armed with proper information, and be prepared to make changes before it's too late. Research shows that post-menopausal women with either benign or malignant ovarian tumors register low progesterone levels prior to surgery.[43] What do all these studies tell us? It seems that the use of progesterone before and after surgery may prepare us for dealing with stress and promote more efficient healing.

CONTEMPORARY SYNTHETIC HORMONES

Recently we have also heard much about a drug called tamoxifen. This is a synthetic hormone, a "nonsteroidal antiestrogen," and is related to the carcinogenic hormone DES. It has been used since 1977 to treat advanced breast cancer but is now believed to have many unusual side effects, some of which are similar to those of estrogen, which can promote more tumors. Nevertheless, doctors are considering tamoxifen for use in hormone replacement therapy. In one study almost half of the tamoxifen recipients "complained of 'persistent vasomotor, gynecologic, or other major side effects.' "[44]

The women enrolled as volunteers in the clinical trials of tamoxifen as a breast cancer deterrent were apparently informed of a statistically known increased risk of endometrial cancer.[45] In February 1996, a review by the International Agency for Research on Cancer—based in Lyon, France and composed of scientists from various countries—concluded definitely, according to published reports, "that there is sufficient evidence to regard tamoxifen as a human carcinogen that increases a woman's risk of developing . . . cancer of the endometrium, the inner lining of the uterus."[46]

Tamoxifen has also been linked to liver cancer, eye disease, and depression. "Despite that," says Dr. Marcus Laux, "there's talk of selling this to millions of women at high risk as a 'preventive'!"[47] No wonder Dr. John Lee says that other countries view our giving tamoxifen to postmenopausal women as "another American joke."[48]

Of course, it's impossible to keep up with all the new synthetic hormones and drugs that continue to be approved for today's market. Appendix B, for example, lists some of the numerous terms that apply to synthetic estrogens and progestins. Certainly, great caution should be used when taking synthetic or conjugated estrogens. One of the most often prescribed, Premarin, is described this way in the physician information sheet: "Premarin is a mixture of estrogensulfates blended to represent the average composition of material derived from pregnant mare's urine (Pre-mar-in). It contains estrone, equilin, 17a-estradiol, equilenin, and 17a-dihydroequilenin."[49] In fact, says Dr. Marcus Laux, "Premarin is the number one best-selling drug in the country,"[50] and contains synthetic additives and "foreign estrogenic elements . . . that we believe [are] dangerous and potentially cancer producing."[51]

So how can some doctors and pharmacists honestly say that Premarin is natural? The authors of *Women on Menopause* write that Premarin "cannot be natural to humans, although it is to horses."[52] We shouldn't be too surprised that we eventually see side effects and even tumor growth as a result of the synthetic components of this drug, not to mention its inevitable contribution to estrogen overload.

And then there are the imitation progesterones, such as the commonly prescribed Provera (methoxyprogesterone acetate). Majid Ali, M.D., calls this progestin a "synthetic molecule that does not fit well with the natural order of hormonal rhythms."[53] Thus, introducing it along with Premarin seems like adding fuel to an already hazardous fire. In fact, in a seminar presentation Norman Shealey, M.D., cautioned that Provera and all the other progestins actually lower DHEA levels, as does Premarin.

When a patient reacts badly to one of these compounds, the drug is put aside and one of the many other synthetic substitutes is prescribed instead. Whenever you have a prescription filled, ask your pharmacist for a copy of the package insert that is prepared for the doctor and the patient. The warnings and contraindications concerning the new drug assigned you will be clearly stated. See appendix B or refer to the *Physicians' Desk Reference* (PDR)[54] and other books available at your library or bookstores.

THE WARNINGS ABOUT SYNTHETIC HORMONES

Women are now paying more attention to the cautions about dangerous side effects. Concerned about their own possible reactions to these drugs, they are looking around for alternatives.

Synthetic estrogen, to begin with, is prescribed by doctors for many situations, such as cramps, acne, irregular periods, birth control, infertility, menopause, postmenopause, and osteoporosis. Precautions come with all drugs, but with the typical estrogen package, the warnings on the inserts are so small that it's easy for your vision to fade following the first few sentences. However, it's essential to read on. Here is an example from the literature for the synthetic hormone Premarin:

> **Endometrial Cancer.** There are reports that if estrogens are used in the postmenopausal period for more than a year, there is increased risk of endometrial cancer (cancer of the lining of the uterus). Women taking estrogens have roughly 5 to 10 times as great a chance of getting this cancer as women who take no estrogens. . . . Estrogens can cause development of other tumors in animals, such as tumors of the breast, cervix, vagina, or liver, when given for a long time.[55]

Estrogen incites cell multiplication, and at high levels it starts activating the formation of cysts and tumors. With regard to the synthetic Premarin, *The People's Pharmacy* tells us that five out of fourteen of the women who are prescribed it for menopausal problems are increasing their risk of cancer of the uterine lining (endometrial carcinoma).[56] And of course, the longer we take estrogen, the higher is our risk for acquiring such cancer.

Progestin therapy together with estrogen is often recommended by your medical doctor, but even so, as the "Information for the Patient" by Mead Johnson Laboratories says:

> The possible risks include unhealthy effects on blood fats (especially a lowering of HDL cholesterol, the "good" blood fat which protects against heart disease risk), unhealthy effects on blood sugar (which might worsen a diabetic condition), and a possible further increase in the breast cancer risk which may be associated with long-term estrogen use.[57]

Even if we do everything right in the way of avoiding harmful radiation and synthetic hormones, the picture is not complete and we are unlikely to succeed in our effort to avoid "diseases of civilization" such as cancer unless we adopt an otherwise healthy lifestyle. Only when our bodies are provided proper nutritional support can substances such as progesterone achieve their full impact on the system. Any approach to cancer prevention or control must embrace the entire person—mind, body, and spirit. We should choose health-care providers who will try to help us reestablish internal balance at deeper levels so that the body can heal itself as it is programmed to do.

A more complete discussion of cancer, nutrition, and mammography is beyond the scope of this book. However, it is vitally important that we become aware of the dangers of blindly following standard allopathic procedures, especially mammograms. It is suggested that the reader refer to the information and selection of books on these subjects in appendix D. In particular, Dr. Ralph Moss's book *Cancer Therapy: The Independent Consumer's Guide to Non-toxic Treatment and Prevention* covers many of these subjects in greater detail.

A GLANCE AT THE FUTURE

Hearing the personal stories of the "incurable" helps us to see clearly the merits of prevention and to understand that the whole truth concerning health is made up of a variety of natural choices. However, it seems evident from diverse studies that more research should be done, not only for women but also for men. To illustrate, *Choices in Healing* describes a case history in which a man with prostate cancer was given all the best conventional procedures, including chemotherapy. Nothing seemed to help, and he became quite weak. However, his doctor prescribed specialized hormone therapy, by means of which he recovered.[58]

This story was interesting but did not leave much of an impression on me until one day when I was having a casual conversation with the owner of our neighborhood health food store concerning the many benefits of natural hormone supplementation for women. During our discussion, a man in the store overheard us and asked, "Can natural progesterone help with male as well as female problems?

The question does need a response, especially in light of statistics published in *The Menopause Industry: How the Medical Establishment Exploits Women:*

> Approximately 132,000 men were diagnosed with prostate cancer in 1992, with a five-year relative survival rate of 77.6% for white men compared to the 79.3% rate for white women with breast cancer.[59]

Author Sandra Coney surmises that the sea of environmental estrogens in which we find ourselves might explain "the rise in male prostate cancer paralleling the rise in female breast cancer."[60]

The best way to answer the question is to quote from a highly respected, popular guidebook entitled *Alternative Medicine: The Definitive Guide,* with chapters written by 350 leading health-care professionals and scientists. The book presents a wealth of effective treatment options. One of the personal accounts, written by a medical doctor, would certainly be of interest to the man who asked the question about men and progesterone:

Topical application of natural progesterone may prove beneficial in the treatment of prostate conditions. The doctor reports working with twelve men, all in their late seventies, who were suffering from osteoporosis. As it has been well established that natural progesterone, applied topically, can relieve osteoporosis, the physician suggested that the men systematically massage it into their skin on a daily basis. All of them began to experience relief from their condition, and later called to tell [the doctor] that, after three months, . . . they were also experiencing an improved urine flow, with less pressure on their prostate glands and noticeable decrease in nightly urination.[61]

The physician who treated these men did not choose to have his name publicized. He preferred to remain anonymous rather than be associated with natural alternatives in healing and then be ridiculed by those of his peers who focus on orthodox medical standards and use drugs to treat their patients. Sadly, this attitude is what creates the greatest obstacle for the many patients who desperately wish to obtain natural products from their medical doctors.

Perhaps when this doctor retires and no longer fears being discriminated against, he will write a book for the large numbers of men who suffer from prostate enlargement and its unpleasant symptoms. The good news that natural progesterone (as also indicated by Dr. Ray Peat[62]) and herbal remedies can actually reverse many prostatic disorders would certainly be a relief to the ten million men in the United States who are afflicted with impotence, of whom only 200,000 actually seek medical help.[63] Indeed, progesterone protection seems to be the missing link in many problems.

Dr. Shealey notes that women after the age of about fifty are not deficient in estrogen but in progesterone. He theorized that if women are deficient in progesterone, this hormone must also decline in men as they age. He could not find any studies in men, only women; so he started using natural progesterone on his older male patients. The progesterone apparently caused one patient's DHEA hormone level to double. His libido soared, and he felt better than he had in years. In fact, Dr. Shealey discovered that the majority of men who used the natural progesterone cream showed a marked increase (60 to 100 percent) in their DHEA.[64] This is important news, since DHEA seems to spark the metabolism of both men and women and helps balance the entire glandular system. More research on the connection between these two hormones is definitely in order.

MELATONIN AND BREAST CANCER

Speaking of hormones, we should mention one that has become a popular item in the newspapers, magazines, talk shows, health stores, and drugstores. Melatonin, a hormone secreted by the pineal gland, has been reported to inhibit cell division and "block the growth of breast tumors in laboratory animals."[65] It is suspected that one's level of melatonin may be an early predictor of some types of cancers, "since women with ER [estrogen-receptor]-positive breast cancer and men with testosterone-dependent prostate cancer have lower levels of melatonin than those without the disease."[66]

Mice given melatonin in their water at night never developed terminal cancer, which often affects the particular breed studied. The mice that didn't get the melatonin lost neuromuscular coordination, their immune and thyroid functions declined, and they died of cancer.[67]

Dr. Ray Peat, when asked about melatonin, cautioned that many of the studies have indeed been done on rats and mice, which are nocturnal animals. He expressed reservations because when melatonin was given instead to pigs, whose hormone system more closely resembles that of humans, it was found to lower progesterone and thyroid hormone levels and to raise estrogen and prolactin.

If you do decide that melatonin is a supplement you would like to try, be aware of possible side effects. The timing of its use seems to be a key factor in preventing estrogen from becoming dangerously high.[68] Researchers have found melatonin injections in the morning to be risky, as given at that time they stimulated tumor growth. However, when given in the evening they retarded cell proliferation.[69]

THE THERAPEUTIC USE OF ESSENTIAL OILS

D. Gary Young, Doctor of Naturopathy, tells us that plant essences known as essential oils have regenerative and immunity-stimulating properties. "Their oxygenating molecules effectively transport nutrients and a myriad of other powerful chemical constituents to the cells," he says, and "bring life to the plants, destroying infections, staving off infestation, aiding in growth, and stimulating healing." These oils are to "plants what blood is to the human body."[70]

These remarkable, life-supporting oils include: lavender (for its ability to heal, relieve stress and inflammation, kill pathogenic microorganisms, etc.), lemon (boosts the immune system, kills bacteria, functions as an antidote for toxins, etc.), peppermint (kills parasites, may be used as an inhalant for respiratory problems, etc.), and ylang-ylang (antiseptic, fights intestinal infections, relieves high blood pressure, etc.). Specifically for hormonal balance, some have found the following essential oils beneficial: bergamot, blue yarrow, clary sage, fennel, and geranium, among others.

As an example of their therapeutic application, Dr. Young's educational tape, *God's Miracle Oils,* quotes from the Associated Press concerning cancer prevention with a combination of soybeans, lavender oil, and orange peel. He refers to the cancer-fighting potential of "small quantities of lavender oil which has been shown to work against breast cancer in live animals."[71] If you'd like to have a better understanding of the healing powers ascribed to essential oils be sure to read *A Reference Guide for Essential Oils* and *An Introduction to Young Living Essential Oils and Aromatherapy* (see appendix E).

THE POWER OF NATURAL ALTERNATIVES

When I began looking into natural means of ridding poisons from my body and counteracting the toxic surroundings that are increasingly difficult to avoid, it was a comfort to become acquainted with like-minded medical doctors who are getting off the track of medicating the symptoms of disease and getting on the track of preventing the cause of disease. My hat goes off to them for thinking of the patient's needs first.

Stuart Berger, M.D., advises that when we choose a medical doctor, we should "avoid powerful ways that our medical system and its illness-centered view of medicine keeps us unhealthy . . . and actually raises our risk of medical problems."[72] Medical doctors, like everyone else in this world, should be held accountable for their actions. However, we also need to be personally accountable by keeping in tune with our bodies and learning about natural substitutes for medication. A drug may be essential in cases of acute disease, but prolonged use of a prescription drug often covers up the underlying cause.

Dr. Lorraine Day faults medical science for attempting to blame cancer and other diseases on genes, thus taking away our personal responsibility. She states—and I can vouch for this from personal experience—that often doctors will tell their patients it's O.K. to eat anything they want![73] She points out that they may know little about nontoxic therapies because they offer less profit potential and do not receive funding for the hugely expensive studies required for approval by the FDA; and the FDA and AMA help determine what insurance companies will and will not pay for.[74]

Let's reflect for a moment on the divergent philosophies. Our distinctive choices in health care may produce entirely different conclusions: (1) Masking the symptoms (temporarily avoiding disease and/or surgery through medication) is focusing on the *disease*. The cause of the illness is still festering—but since medication may slow down the need for immediate surgery, the medical field calls this prevention. (2) Aiding the body to avoid disease or to heal itself through good nutrition and natural therapies is focusing on health.

We should recognize abuses. If some of our gynecologists could have a taste of their own medicine (unnecessary needle biopsies, hysterectomies, mammograms, toxic synthetic estrogens or progestins, tamoxifen, etc.), they might be more inclined to respect women's rights to natural choices and would give higher priority to knowing what will harm the body and what will heal. It could be that we—as patients who have to live with illnesses and the side effects of drug treatment—need to be the ones to make a change: to become fit in order to be survivors. It is up to us to reshape the public's attitude toward health and disease.

With a little curiosity and effort we, too, can learn not only how to avoid the hazards of certain modern technology but also how to counter the depletion of our energy by the continual use of toxic drugs. The suffering that results today from the many drug-driven diseases is often unnecessary. We deserve more than this, and so do our families.

A phenomenal healing potential exists within the human body. Once released, the life force can manifest itself in a degree of wellness that many of us don't even realize possible. Admittedly, learning how to harness this on our own can be time-consuming and trying. But once we find the right professional to work with us, we'll discover that *implementing what we instinctively know to be true is truly living*. By striving to achieve a harmonious synchronization of the body's systems, we'll enable the many parts to work together to create a healthier, disease-defying whole.

For specific advice on natural hormone replacement therapy, please see chapter 7. It tells you how to make wise choices, understand the doctor–patient relationship, and confidently communicate your needs. Self-help advice and valuable resources, including names of recommended books, pharmacies, and manufacturers, are found in appendix F (Information for Your Doctor) and appendix G (Sources of Natural Progesterone).

PART III

Making
Assertive
Lifestyle
Changes

CHAPTER 6

HORMONAL SUPPORT
FROM OUR FOODS

A person should take every care of his body . . . not for the sake of the
body, but in order that the soul in a sound body may act . . . rightly, and
may have the body as an organ perfectly obedient to it.
                                               Emanuel Swedenborg[1]

To use natural healing methods effectively requires modifying our habits and developing a good deal of discipline. At one time I had trouble making such a transition in my own health regimen. Instead of getting involved with prevention, I tried to tell myself that I was not going to have menopausal problems. If I didn't think about it and just kept busy, I hoped to bypass "the change." So instead of facing it and looking for alternative solutions, I allowed my symptoms to grow worse. The stress and the aging process simply accelerated. Not in control of my life, and skeptical of natural remedies, I was paying very dearly for my reluctance to deal with the transition. It was from this low point that I gradually became more aware of prevention and the necessity of a healthy lifestyle.

Women suffering from intense loss of energy or drive during PMS will be interested in Dr. Katharina Dalton's recommendation that they snack on a complex carbohydrate food every three hours. In fact, many nutritionists "encourage sufferers to be 'grazers,' eating at frequent intervals, and not 'gorgers', only eating three times a day," and point out that starchy foods such as rice, oats, potato, rye, or corn prevent "adrenaline release from low blood glucose."[2,*]

---

* This would not apply to those concerned about a tendency to diabetes. Raymond Peat, Ph.D. recommends a *low*-starch diet to avoid elevated blood sugar, insulin and fat over-production, cell death, and allergies.[3]

It's quite interesting that some of these same foods are discussed in *Modern Pharmacognosy* as sources of progesterone: "Small amounts are . . . found in plants, e.g., yeast, rice, wheat, cabbage, and potato."[4] Dr. John Lee says that in "cultures whose diets are rich in fresh vegetables of all sorts, progesterone deficiency does not exist. . . . Many (over 5,000 known) plants make sterols that are progestogenic."[5] As for estrogen activity, it is associated with soy products, such as tofu, soybeans, and soy milk, as well as alfalfa, licorice, pomegranates, and boiled beans.[6, 7] Those wishing to limit their total estrogen intake may find such information helpful. Fermented soy items such as miso and soy sauce provide less potent levels of estrogen.

American and Finnish researchers report that "Japanese women who eat about 2 ounces of soybean foods a day have less severe menopausal symptoms than Western women. The active ingredient, isoflavonoid, boosts estrogenic activity in soybeans."[8] Asian women, who eat soybeans, tofu (bean curd), and miso (soybean paste), also have low rates of heart disease, cancer, and osteoporosis. Although it has been theorized that there may be a connection,[9] many different dietary and lifestyle factors are likely involved.

Dr. John Lee offers the explanation that the "phytoestrogens" made by plants such as soybeans may be weaker than the estrogens made in the human body but have similar estrogen-like effects. He says that they occupy a woman's receptor sites and thus may protect her from an excess of her own more potent estrogens, which might otherwise overstimulate her tissues and promote cancerous changes. When they fill those sites, says Dr. Julian Whitaker, "the harmful forms of estrogen have no place to bind, and therefore cannot exert their cancer-promoting mechanisms."[10]

Dr. Marcus Laux, referring to the work of a leader in the field of phytoestrogen research, gives a somewhat similar view of the reason soy may be of value to the menopausal woman. Dr. Herman Adlercreutz found that "the estrogenic effects of soy foods can be attributed to the isoflavones (phytoestrogens) they contain. When you eat soy foods regularly, you have high levels of isoflavones in your body. These phytoestrogens in soy bind to estrogen receptors that would otherwise remain empty because your body's production of estrogen has dropped."[11]

WHICH SOY FOODS SHOULD WE CHOOSE?

As individuals with varying needs, we need to be able to make educated decisions regarding health and diet. An excellent treatise on the subject of soy as a food source appeared in *Health Freedom News*. Its argument is that only *fermented* soy products such as miso, tempeh, natto, and soy or tamari sauce are digestible or safe, because fermentation, to some degree, reduces soybeans' powerful enzyme inhibitors (which interfere with protein digestion), clot-promoting agents, and phytic acid. Phytates block the intestinal uptake of essential minerals, especially zinc, more in soy than in any other grain or legume. "Precipitated" soy products such as tofu especially have this effect if they are not accompanied by a little fish or meat protein.[12] It might thus be wise to concentrate mainly on fermented products.

Much of the difficulty with nonfermented soy foods comes from their processing. Textured vegetable protein, the result of high-temperature, high-pressure extrusion, has been shown to damage vital organs in animals. The fatty acids in soybeans are very susceptible to rancidity under high pressure and heat; and soybean oil cannot be extracted without dangerous solvents, some of which remain in the final product.[13] Soybeans are also listed in *Nutrition for Women* as one of the foods that inhibit thyroid function.[14]

Soy milk can be highly allergenic in itself, lacking in needed natural vitamins and cholesterol. Its production creates an incomplete, denatured protein with possible chemical contamination. In addition, soy baby formula contains many additives,[15] and some researchers stress the need to study the question of feeding infants and children, sometimes for years, a substance known to have estrogenic properties.[16] The most natural solution? Breastfeeding, enhanced by progesterone therapy. However, if this is not possible, homemade formulas can be prepared with a base of broth or raw, organic whole milk (see appendix C).

FRIENDLY HERBS FOR OUR HORMONES

The plant world provides us both food and medicine. Herbs come in many forms, from teas to capsules and everything in between, and can be used for most every condition and symptom. Many books have been written informing us how much to take and for which illness. We will very briefly review some of those herbs that are for menopausal problems. It is important to remember, however, that herbs are powerful and some can be dangerous if combined with certain prescription drugs or taken in excess. Always consult a reputable herbalist before deciding which herbs are right for you.

Some dear friends gave me an herb book that notes the natural estrogenic qualities of damiana for relief of hot flashes, and blessed thistle and saw palmetto for other female disorders.[17] Marcus Laux, N.D., expands the list of plants with estrogen-receptor activity: dong quai, alfalfa, black cohosh, fennel, anise, flaxseed, yellow dock, hops, sesame, garlic, many kinds of fresh and dried beans, cabbage, and olives.[18]

The Herb Lady's Notebook stresses that progesterone is the most important hormone for balancing the thyroid and entire glandular system.[19] Especially affected is adrenal activity.[20] Herbal progesterone-like qualities are found in sarsaparilla, ginseng, licorice root, bloodroot, red clover, mandrake, nettle leaf, nutmeg, damiana, turmeric, sage, oregano, thyme, and unicorn root.[21] Dr. John Lee points out that dong quai and fennel have active progestogenic as well as estrogenic substances.[22]

Dr. Julian Whitaker adds that motherwort, chamomile, and chasteberry are good for relaxing the nervous system and reducing hot flashes, night sweats, and heart palpitations. In addition, oat straw and blue and black cohosh are good for muscular toning and as uterine tonics. For muscular spasms or cramps, lack of energy, or mild depression, skullcap is recommended. To relieve vaginal dryness you could include licorice, wild yam, and goldenseal. There are also topical lubricants, which include calendula (marigold) and vitamin E.[23]

Jesse Hanley, M.D., notes that because women are not getting the help they want from mainstream medicine, they are experimenting with self-help treatment through herbs. Dr. Hanley warns that although certain estrogenic herbs might relieve your symptoms, they could also increase the risk of fibroids, tumors, and cancers, and therefore advises us, "Make sure your formulas contain progesterone-precursor herbs, such as sarsaparilla and chaste berry *(Vitex agnus cassus)*."[24] (Sarsaparilla, incidentally, is also a precursor of testosterone.) James Jamieson, pharmacologist, also emphasizes the importance of balancing the estrogenic herbs with a natural progesterone source.

THE B-COMPLEX VITAMINS

The significance of the B vitamins became clear to me when I read Dr. Carlton Fredericks' findings that B-complex will help to convert estrogens, such as estradiol, to estriol for excretion from the body. This may be an even more important issue for women who are considered at risk for breast cancer. As Dr. Richard Passwater tells us, "Breast cancer patients have lower levels of estriol than normal" (see chapter 5). Therefore, including B-complex on a daily basis from adolescence on may actively assist in this conversion process.[25]

This was confirmed by Dr. Fredericks, who found that cancer is linked to deficiencies in B-complex (especially folic acid) along with vitamins C, E, and A. "B-complex vitamins help the liver regulate the estrogen level," says Dr. Fredericks.[26] He goes on to advise that the "control of estrogen is a nutritional process. . . . It is not properly achieved by prescribing more estrogen."[27] He says that "vitamin B complex cuts down excess activity of estrogen, whether prescribed for menopause or birth control or produced by the woman's own ovaries. . . . [It] helps to counteract the effects of high levels of estrogen which has not been detoxified by the liver."[28]

When taking a B vitamin, remember that they all work together, each one needing the other. In general, almost all of us need a B-complex supplement because these vitamins are often lost in the processing of foods. Dr. Fredericks states in his book *Guide to Women's Nutrition: Dietary Advice for Women of All Ages* that progesterone can be stimulated by vitamin B_6, which can be useful *in moderate amounts* for women who suffer from bloating and water retention. He says that if "a woman takes the birth control pill, she develops vitamin B_6 deficiency and then folic acid deficiency."[29] Thus, he says, B_6 may counter "the negative effects of the birth control pill. . . . Vitamin B_6 has also been helpful for women in offsetting the tendency to store water, particularly in the premenstrual week, and helps clear up premenstrual acne."[30] He also states that you can encourage "safe progesterone production with vitamin B_6—a critical factor in controlling estrogen."

Furthermore, according to Gail Sheehy in *The Silent Passage*, taking progesterone with B_6 may be helpful for women suffering from exhaustion.[31] As for some of the foods that contain B_6, Dr. Fredericks lists "beef liver, kidney, fresh fish, bananas, cabbage, avocados, peanuts, walnuts, raisins, prunes, and cereal grains."[32]

A NATURAL APPROACH TO DEALING WITH ANEMIA

In some situations, iron pills may intensify anemia "or even be the cause of it," warns Dr. Ray Peat. He points out also that it "would seem reasonable to consider the role of vitamin E in anemia, before giving a woman iron pills."[33] Concentrating on one's diet is safer than taking iron salts, which have a destructive effect on vitamin E. Iron accumulation can lead to inflammation, calcification, cancer, and suppression of progesterone synthesis.[34, 35] Should you have a deficiency, natural sources of iron include red meat, liver, eggs, wheat bran, wheat germ, oatmeal, brown rice, raisins, dried apricots, prunes, dates, apples, bananas, cherries, grapes, oranges, parsley, carrots, celery, onions, and blackstrap molasses.

Dr. Peat also makes the interesting observation that more often than not, anemia is a thyroid/estrogen problem, related directly to the temperature of the long bones of the arms and legs, where red blood cells are formed in the bone marrow. Thus, in addition to warm clothing, an answer may be support for the thyroid, which is instrumental in body temperature regulation.

VITAMINS A AND E AND PROGESTERONE: A DYNAMIC TRIO

Dr. Fredericks' *Guide to Women's Nutrition* gives some information that you can relay to your medical doctor, should he or she be skeptical about the use of vitamin E for PMS problems. Tell him or her that the *Journal of the American Medical Association* (which, as Dr. Fredericks says, "the orthodox physician is likely to regard with some awe if not trepidation"[36]) states that vitamin E in doses ranging from 150 to 600 units daily decreases menstrual flareups, anxiety, moodiness, irritability, headaches, fatigue, pounding of the heart, and dizziness.[37]

If natural progesterone is not readily available in an emergency, there are vitamins we can temporarily use that also help to balance estrogen levels, especially vitamins A and E. In fact, Dr. Ray Peat says that vitamin A and pantothenic acid (B_5) "promote natural progesterone synthesis,"[38] adding that both vitamin E and progesterone increase the amount of oxygen in the uterus.[39, 40]

Vitamin E has relieved some women of a variety of menopausal conditions, such as hot flashes and emotional problems.[41] In the book *Cancer and Its Nutritional Therapies*, Dr. Richard Passwater informs us that "Vitamin E has been used successfully by many physicians to make fibrous cysts disappear."[42]

As for vitamin A, researchers recognized the relationship between vitamin A deficiency and cancer as far back as 1925. *Cancer and Its Nutritional Therapies* refers to studies by Dr. Harold Manner of Loyola University in Chicago, wherein "vitamin A was a critical component of his protocol to *cure* breast cancer in mice." Dr. Manner also cites laboratory evidence that "cancers such as breast, lung, and skin tumors can be cured by treatment with vitamin A." This vitamin reversed "the effects of carcinogens in tissue cultures from the prostates of mice."[43] Dr. Ray Peat corroborates that "vitamin A protects some tissues, such as the breasts, against estrogen's effects, including cancer, and generally offers protection against estrogen by increasing progesterone."[44]

Vitamin E and riboflavin (B_2 vitamin needed in red blood formation) will provide relief from hot flashes and severe PMS problems, says Dr. Fredricks; and a cream combining vitamins E and A has been successful for a large number of women with vaginal atrophy.[45]

PURE WATER: THE BEST MEDICINE

The importance of consuming plenty of water is emphasized in the book *Your Body's Many Cries for Water*. Its author, Dr. Batmanghelidj, scientifically explains why major diseases are often simply due to chronic, long-term dehydration. The water you drink will bring about cell volume balance in your body and result in more efficient cell activity.[46] When the body is in a hydrated state, proteins and enzymes become much more efficient. Dr. Batmanghelidj tells us, "The damage occurs at a level of persistent dehydration that does not necessarily demonstrate a 'dry mouth signal.'" Thus if we only drink when we are thirsty, our cells have already suffered.[47]

Of particular and critical interest to women is the role of chronic dehydration in the development of breast cancer. Dr. Batmanghelidj says the stress it creates "will increase the secretion of a hormone called prolactin, which can at times cause the breast to transform into cancerous tissue. . . . Also, the dehydration would alter the balance of amino acids, and allow more DNA errors during cell division."[48] Dr. Batmanghelidj says dehydration also suppresses the immune system by making one's natural killer cells less active.

He explains, "The breast is a water-secreting organ. . . . Whether you are having a child or not makes no difference. The breast must be ready to fulfill its predestined role. . . . If a woman already has breast cancer, drinking plenty of water," he says, would assist with any therapy by flushing out toxins. "If you do not have breast cancer, or want to prevent a metastasis from occurring, it is urgent that you drink enough water. If you don't, your breast may suffer horribly because of its unique role in supplying fluids."[49]

The doctor makes the point that water "improves the uptake of hormones by the cells" by its action on the receptors at the cell membrane. He further stresses the necessity of avoiding caffeinated beverages or compensating for their diuretic action, and also the necessity of adequate salt intake in order to retain sufficient water in one's cells.[50]

THE BODY'S NEED FOR GOOD CHOLESTEROL

Let's hear it for the good cholesterol that's in our bodies. This cholesterol (which Dr. John Lee says is of concern mainly when it is *oxidized*) is actually the precursor of progesterone, estrogen, DHEA, and other hormones. Wallace L. Simons, R.Ph., of Women's International Pharmacy, says that we've been sold a real bill of goods about cholesterol. As long as we consume plenty of antioxidants, we should not have a problem;

arteriosclerosis is brought on in part when the "bad fats" are allowed to break down into free radicals.

We are told to stay away from many foods in order to avoid cholesterol problems; however, we may be mistakenly steered away from important foods in order to lower our dietary fat content. An example of this is seen in an interesting study from Denmark concerning the inclusion of eggs in our diet. Dr. Earl Mindell provides us with some vital facts: "When 21 healthy adults ranging in ages [from] 23 to 52 years old were given two boiled (not fried) eggs every day plus their ordinary diet, their *good* blood cholesterol went up 10 percent, while their total blood cholesterol only went up 4 percent after six weeks." It seems that moderate egg consumption may be quite beneficial where cholesterol problems are concerned.[51] Of great interest to us in this discussion of hormones is the statement by Dr. Ray Peat, in his book *Nutrition for Women*, that including eggs and liver in the diet will promote the formation of progesterone.[52]

We should also consider butter a preferred choice over margarine, an artificially engineered product whose partially hardened oils can be toxic and damaging to the arteries and the immune system. The vitamins A and D that butter contains "are essential to the proper absorption of calcium and hence necessary for strong bones and teeth."[53] Butter is also an excellent source of a highly absorbable form of iodine and assists the proper functioning of the thyroid gland.[54] Its lecithin helps to assimilate and metabolize fat constituents such as cholesterol,[55] and it is rich in antioxidants, including selenium and vitamins A and E. Even cholesterol is an antioxidant. Both vitamin A and the short- and medium-chain fatty acids found in butter are valuable to the immune system, and its lipids protect gastrointestinal health. Because these fatty acids are burned for quick energy rather than stored in the fat tissue, *Health Freedom News* states that "the notion that butter causes weight gain is a sad misconception."[56]

HORMONES, FATTY ACIDS, AND LIPOPROTEINS: SORTING OUT THE ISSUES

Clearly, the whole question of fats in our diet is one we need to approach cautiously. Fad follows upon fad, and we are at the mercy of public relations teams and pop journalists who often fail to delve into all the scientific facts. True, the subject is important, as it relates not only to weight control, heart disease, and cancer but also to the formation of all our steroid hormones, including progesterone.

However, there are numerous interesting and ongoing studies currently under way regarding fats and oils and their effect on our health, and not all the facts are in.

Dr. Ray Peat has made a comprehensive study of, as he puts it, "the whole history of research on the biological effects of dietary fats."[57] He writes that when the LDL cholesterol (the supposedly bad kind) is too low, the body may be hindered from producing enough progesterone.[58]

On the basis of his extensive research, Dr. Peat challenges some of the modern-day assumptions about the "essential fatty acids" that he says originated with faulty interpretation of certain tests performed on rats in the 1930s. In fact, he says, "it is now known that polyunsaturated fats interfere with thyroid hormone in just about every conceivable way"[59]—and they have a strong estrogen-promoting action.[60] He emphasizes that excessive consumption of these fats leads directly to the development of such degenerative scourges as cancer, diabetes, heart disease, arthritis, osteoporosis, and connective tissue disease.

But what is equally startling, and exciting because it offers a way to reverse these frightening effects, is that Dr. Peat has discovered much-maligned coconut oil to be part of the solution. This product, unfortunately, is currently out of favor, again because of unscientific studies and inaccurate "bad press."

Explains Dr. Peat, "The easily oxidized short- and medium-chain saturated fatty acids of coconut oil provide a source of energy that protects our tissues against the toxic inhibitory effects of the unsaturated fatty acids and reduces their anti-thyroid effects." This oil contains immunity-boosting lauric acid, also found in mother's milk. For these reasons, its regular use offers protection against disease and premature aging. Dr. Peat recommends about an ounce a day. One way to use it is as a salad dressing, mixed with olive oil, cider vinegar, and salt.[61, 62]

An added bonus is that by increasing the metabolic rate via the thyroid, coconut oil, in spite of being a fat, has been known to bring about an amazing loss of excess weight. Farmers, in fact, who thought it would be an inexpensive way to fatten their animals, found it had just the opposite effect![63]

In certain areas, grocery wholesalers may be able to get coconut oil for you quite inexpensively. Ask for "76-degree melt," and you will be getting a natural product which is better than the refined version sold for external use in some health stores. Another source is certain herb shops or ethnic markets and restaurants. Dr. Peat is also making available a high-quality product through his organization, Kenogen (see appendix G).

As for the other oils—we have a right to feel confused! Some, of course, are better than others. In particular, the omega-3 fatty acids (found in deep-water fish such as salmon, mackerel, tuna, menhaden, herring, cod, haddock, and sardines)[64, 65] and the oils containing a high ratio of monounsaturates (extra-virgin olive, canola, flaxseed, and high-oleic safflower or sunflower) have gotten the best publicity for their supposed ability to enhance immune function and reduce LDL cholesterol.[66] In the case of some of these, however, the debate continues. You will also hear about the "essential" fatty acids in oils such as cod liver, evening primrose, and black currant, as well as in such food sources as walnuts, pumpkin seeds, and avocados.[67]

You may have read promising things about flaxseed oil and its use in cancer treatment. It will be wonderful if these things turn out to be true. But just how much of the information has been promulgated by manufacturers of flaxseed products? Dr. Paavo Airola warned about the strong tendency of this oil to rancidity,[68] and Dr. Peat considers it the most carcinogenic of oils!

So if you do choose to include these items in your diet, consume them only when they are very fresh. Refrigeration is recommended, and they should be protected from light. Also, take plenty of antioxidants, especially vitamin E, to counter any possible adverse effects. Be aware of the controversy, and use moderation in all things—we still have a lot to learn.

THE FIGHT FOR NATURAL HEALING— A BATTLE YOU CAN'T LOSE

The more we begin to understand what the body needs, the more we can support its innate design for utilizing natural foods to cleanse and regenerate tissue. By choosing from the many health-care approaches those that meet our individual requirements, we can transform infirmity into vitality. So, before considering yet another drug for a condition that is still festering, why not take advantage of the unadulterated forms of healing that are gentle, yet powerful in addressing the cause of disease? Nature already offers us curative substances that will help us achieve optimum energy and make us less vulnerable to the effects of environmental stress. Making such an important personal effort will influence much more than just our physical health.

CHAPTER 7

PLANNING A
PERSONAL APPROACH

*You need to be the instigators. . . . You cannot expect your
doctor to assume control and solve all your problems—
M.D. does not stand for Medical Deity.*
                                   Christiane Northrup, M.D.[1]

The best way to avoid the "midlife crisis" experienced by many women
may well be to take a saner approach to health. This involves learning about
natural therapies, unknown to so many, that will restore the body's internal
balance. Looking back over the years I wonder why I didn't take time to
educate myself sooner about the many forms of alternative health care. In
justifying my case, I reflect on how much easier it was just to let the medical
doctor make all the decisions. How easy it is to become dependent on the
advice of others, especially when we feel sick and tired! Unfortunately,
that's not a wise attitude, as many of us can attest in hindsight.[2, 3]

Being unaware of our options, however, we accept, largely on faith alone,
the standard medical treatment currently in vogue. Unfortunately, the relief
it offers may be only temporary. Worse, its multiple side effects often force
us to deal with future disorders that are routinely treated with biopsies,
mastectomies, hysterectomies, and drugs. We go through years of confusion, wondering where to turn.

In retrospect, I realized the imprudence of being completely dependent on standard medical treatment (in this case, synthetic HRT) when
I was given a different diagnosis by every doctor I visited. There
seemed to be as many points of view as there were doctors. I learned
eventually that their information about drugs comes largely from the
pharmaceutical companies and, in turn, the sales force that is promoting specific products. Down the road the patient's hope for a cure will

hinge upon which drug or hormone was being marketed when she received her prescription.

Some doctors are taught that estrogen is the only hormone that's necessary for the treatment of osteoporosis. And when the patient asks about the warnings listed on the estrogen (or progestin) packet, the advice she receives is this: "Not to worry—remember, there are more benefits than risks when using this hormone!!" After my appointment with the particular gynecologist, internist, or endocrinologist of the day, I would always return home with the same kind of frustrations and fears, wondering why the subject of risks was never fully discussed.

BLIND ACCEPTANCE

My sole reliance on this system kept me at a low ebb during my early menopausal years. However, this was nothing compared with the ordeal of the women I knew who had been on synthetic hormones for several decades. Initially, some of these women had felt tremendous relief when estrogen was prescribed. However, as time went on, complications began to surface. Nevertheless, they decided to endure the fluid retention, weight gain, and other problems with the thought that it was better than trying to tolerate the menopausal symptoms. Too bad, because these signals in response to HRT are the body's way of crying out for us to STOP—and for very good reasons!

Nevertheless, many of us continue to consult orthodox medical practitioners. Although the process is expensive and time-consuming, we go along with it because we need help. One has to wonder how long this system of dependence, chaos, and sometimes intimidation will persist.

Expressing many women's discouragement with synthetic HRT, Gail Sheehy refers to Germaine Greer, who "properly faults the medical establishment for conspiring to make women dependent on pills and patches that have been woefully undertested."[4] And we ask, "Why is this, when medical research is supposed to be so infallible?" The answer may lie in an examination of where the profits are coming from. As we have already demonstrated, a synthesized commercial compound that requires a doctor's prescription for its sale commands a much higher market value—not to mention that it will not have an identical action to that of its natural counterpart and will often create toxic responses.[5, 6] The warnings in the *Physicians' Desk Reference* list the many unwanted and often serious side effects. But unaware of any of this, we may come home from a medical doctor's office exasperated once again, feeling more and more frustrated.

I see a real need for women to be more alert to the barriers and sometimes even incompetence that stand in the way of our health. As we seek that path less traveled, more and more of us will realize that the remedy we have been waiting for has already been found and is presently being utilized by many doctors and their patients. This information was discovered as far back as the early 1930s, so it saddens me that all women today are not informed about NHRT. We should be given the chance to *first* try a natural product that has been shown, even proven,[7-12] to relieve persistent afflictions without side effects. Instead, synthetic compounds are prescribed and so often complicate our already stressful lives.

Hysterectomy . . . endometriosis . . . osteoporosis . . . breast, cervical, or uterine cancer—these no longer have to be specters overshadowing our lives. Dis-ease of the body, mind, and spirit can be a thing of the past as we begin to focus on prevention: proper diet, exercise, supplemental natural hormones. Once we experience for ourselves the impressive results, we'll be zealous to spread the good news about NHRT and to fight for our health and our well-being.

By listening to the wisdom of our own bodies and taking advantage of the evaluation tools that are available to us, we can learn whether there are deficiencies or excesses in our hormone levels and thus avoid the ordeals just mentioned. However, not everyone has the time, energy, or funds to have expensive tests done at a medical specialist's office. Just thinking about the difficulty of finding a doctor who understands the difference between the natural and the synthetic—not to mention trying to find time in an already busy and stressed-filled agenda to schedule an appointment— can be discouraging.

DIAGNOSTIC TESTING: WHAT, WHERE, WHY

Here are some guidelines to make the transition to a more self-directed program easier for you. Before you decide to have your hormone levels tested, let's first explain the difference between blood and saliva tests. The traditional means of testing hormone levels has been by serum blood test. With the recent focus on progesterone, however, it is now understood that only 1 to 9 percent of the progesterone found in blood serum is biologically active. However, this test indentifies mainly hormones made internally by the body, which are bound in a special protein coating that enables them to be carried in the blood. These bound hormones are not biologically active. Their important component is the level of unbound hormone, generally less than 10 percent in blood serum. Progesterone absorbed through the skin, on the other hand, is not bound in this type of coating. It is fully active, and while it may take a while to appear in the blood serum, it appears quickly in the saliva where it confirms that the progesterone has been absorbed and is available for use by the body. As a result a saliva test is a more convenient and efficient way for the doctor to test for any possible progesterone deficiency.[13]

Fortunately, companies such as Aeron LifeCycles [(800) 631 7900] make test boxes that can be sent to you for use in your own home. You can perform a convenient and noninvasive saliva test (no needles) that will measure any or all of the following hormones: estradiol, progesterone, DHEA, and testosterone. Dr. Debbie Moskowitz says that this simple procedure "gives women the ability to monitor changes associated with various alternative approaches, such as diet, herbs, or creams [and] get information regarding baseline hormone levels, or to test hormone levels during a certain phase of the menstrual cycle."[14] For more information, call Transitions for Health at (800) 888-6814.

You can also ask your doctor about the comprehensive hormone balance evaluations that are performed by Diagnos-Techs, Inc. on the basis of saliva specimens collected at the patient's convenience. Kits are supplied to practitioners such as medical doctors, chiropractors, naturopaths, acupuncturists, and any other licensed doctors at no cost; your samples are sent to their lab. Diagnos-Techs will determine your baseline levels and evaluate the degree of deficiency in any of the hormones (estradiol, estriol, progesterone, testosterone, or DHEA). They will then report to your doctor the type and amount of hormonal support you may require. For this service, contact Diagnos-Techs, Inc., 6620 South 192nd Place, J-104, Kent, Washington 98032, or call (206) 251-0596.

The decision as to whether to have a hormone saliva test or not often brings up many questions. A lot of guesswork still seems to be involved in the administration and interpretation of the assays. Some doctors tell you not to take hormones for a couple of weeks prior to taking the saliva test, whereas others say eight to forty-eight hours and some tell you to take the test along with your normal hormone replacement program. Furthermore, you may be told to pick a day when your stress levels will not be high (I'd like to meet the woman who can orchestrate such a day). Also, one's hormones can fluctuate quite a bit, and what's normal for one woman is not necessarily so for another.[15] These levels can even change from one month to another, depending on many factors.

With all of these variables, I wonder if saliva tests can truly be a reliable reflection of our daily hormonal status. The most accurate tool may be just listening to one's own body, turning to the healer within, and then taking steps to ensure that a natural balance of hormones is achieved. As Dr. Marcus Laux states,

> Many internists and gynecologists put women on HRT drugs without testing their hormone levels. Some just test a woman's LH and FSH levels. Although we recommend getting a full testing of all hormone levels—a full steroid panel, as it's called—if your doctor is unable to interpret the results, it won't be that helpful, and *the best measure will be how you feel.*[16] (Emphasis added)

Such a steroid panel is available in the form of a urine test that can be ordered by your doctor. This test, called the twenty-four-hour urine sex hormone profile, measures the average output of various hormones throughout the day. The continuous collection of your urine during a twenty-four-hour period avoids the inaccuracies of test results based on only one reading. Doctors interested in more information may call Meridian Valley Clinical Lab at (253) 859-8700.

Along with these specific tests, a baseline gynecological exam is important in ruling out any underlying pathology as the cause of your premenstrual, menopausal, or postmenopausal symptoms. Once you receive a clean slate as far as more serious disease is concerned, you can turn to a preventive program of hormone replacement. The key here is to find a physician supportive of the natural, nonsynthetic, noninvasive approach that recognizes that *dis-ease* and many female problems stem from an absence of homeostasis (inner balance).

As for testing for bone mineral density, the lowest radiation dose is produced by DEXA screening, which also allows for a high degree of precision and accuracy (within 1 to 2 percent on tests of the lateral lumbar spine). These tests are available in most major cities throughout the United States. For the location closest to you, contact the National Osteoporosis Foundation at (800) 464-6700.

Another resource is the OsteoGram Analysis Center, which uses radiographic absorptiometry technology and computerized analysis of standard hand X rays for measurement of the bone mass of the middle fingers. The analysis will show a patient's bone density measured in terms of a *T*-score (number of standard deviations above or below the reference bone mass for normal young adults). The company provides a starter kit for converting any standard X-ray unit into a bone densitometer. Ask your doctor to phone (800) 806-5639, or write for an application: P.O. Box 47058, Gardena, California 90248.

The Great Smokies Diagnostic Lab [(800) 652-4762)] offers a service for women wishing to monitor the effectiveness of their treatment relative to possible ongoing bone density loss. The lab uses a simple, noninvasive urine collection system to diagnose candidates for antiresorption therapy. The recently developed bone resorption assay performed by MetaMetrix Medical Laboratory also involves a simple urine test that identifies accelerated demineralization. It measures the collagen fragment in the urine that is thought to be specific to the resorption process and an indicator of progressive bone breakdown. According to the laboratory, "Recent research has demonstrated the clinical utility, occurrence, and sensitivity of this procedure." You can contact them at (800) 221-4640.

By using these resources, growing numbers of women are taking responsibility for their own health and finding a sense of confidence, worth, and freedom. They are weaning themselves away from their traditional dependence on and continual use of synthetic drugs and no longer accepting these prescriptions unquestioningly. Perhaps some day, with the trend toward self-monitoring, natural therapies, and personal initiative, the prescribing of synthetic hormones to women will be labeled malpractice, and the prescribing of any form of estrogen unopposed by natural progesterone will be a violation of insurance codes.

WISER CHOICES

Better times are just around the corner. But even as it becomes easier to find that special doctor who is familiar with the latest research on plant-based hormones, we need to understand that there are as many viewpoints and outlooks as there are physicians when it comes to drugs versus natural hormones.

Once you have personally delved into and understood the various roles played by natural progesterone in the metabolic and physiologic operations of the body, you can better discuss your health needs with your physician. You alone can feel the good or bad signals the body communicates, and you are the one who has to live with any side effects. How empowering to learn about this on your own!

So when faced with having to decide between prescribed medication and natural solutions, stand by your convictions and common sense. Be prepared to be calm when approaching doctors who have spent years learning about drug therapy. I have found that some do not appreciate being told of treatments they have not been schooled in.

A Case of Intimidation

One of my friends went to great lengths to research, read, and copy scientific literature from medical journals regarding natural substitutes for synthetic hormones, and then presented the information to her doctor. She even took him a book on the subject, written by a medical doctor who had been using a natural progesterone cream for over twenty years with great success. Her doctor responded to all her efforts with this emphatic statement: "I'm not convinced! I will continue treating you according to *conventional medical standards.*" Women face this very attitude every day, and it leaves very little, if any, room for further discussion.

This form of belittling, however subtle, is not helpful, nor is it respectful. The patient I have just described is a nurse. She is an intelligent, well-respected woman, known for her many useful endeavors in the community. She is also interested in obtaining the very best health care for herself and her loved ones. Her experience could have made her unsure of herself and unwilling to pursue natural HRT any further. However, she did follow through in spite of her doctor's attitude and now feels better for it.

If you have had a similar experience, you need to keep in mind that there are always other health-care providers to turn to—doctors who do not allow their egos to influence their judgment about the best treatment for their patients, and who will at least look into natural therapies that have proved to be effective in the long run. These are the ones we need to seek out to be our family physicians, and the ones we need to tell our friends about. Wise counselors with this propensity, in fact, are quoted often in this book. We cannot afford anything less than to be treated by a doctor who is informed in the most important areas of natural health care.

Communicating Your Needs

When my gynecologist received the information I had mailed to her (see appendix F), she actually called me on the phone herself. She thanked me for collecting so much research and sharing all the literature with her. She told me she would not only use this information herself but make it available to her other patients. She then called in a prescription for me for natural progesterone and estriol. Needless to say, I was delighted at her response and relieved that I was finally going to receive the desperately needed natural hormones.

Likewise, a friend who has been supplying her open-minded physician with literature was told, "I'm glad you're my patient. It's really made me look at natural hormones!" So the more that we, as patients, continue to speak out, the more the medical profession will begin to acknowledge the importance of natural progesterone over artificial drugs. An article under the auspices of ten M.D.s and eight Ph.D.s illustrates this in roundabout fashion, conceding: *"Oral micronized progesterone, a natural form of the hormone, may provoke fewer side effects than its synthetic cousins."* [17]

So encourage your own doctor to read the information from the *Journal of the American Medical Association* [18] and *Medical Hypotheses* [19] regarding the body's need for nonsynthetic hormone replacement. When the doctor reflects on the data, he or she will then recognize that such treatment is not only desirable but also endorsed by medical authorities—endorsed years ago when facts were not compromised or influenced by profits.

So forge ahead! And in your search for "Dr. Good," remember to be positive, be assertive, and don't forget there are other good doctors out there waiting to serve you. They will be quite excited about what you have found and will respect you for going the extra mile. When you find one who really listens and is open to a higher level of doctor–patient communication, you will feel as if you have found not only a doctor but a friend. And as doctor and friend she would agree, "The more you know, the healthier you will be."

Most physicians are well aware that synthetic HRT can create uncomfortable and dangerous side effects, so if a natural product has been observed to have remarkable results with other patients, they may be quite interested in introducing it in their practices. Accordingly, appendix F provides a suggested form letter, to be changed as fits your needs. Feel free to copy the accompanying pages and give them to your doctor. This will allow her to examine a sampling of the extensive research already completed and available, should she wish to pursue the subject further.

Sending a letter or calling the office first is more efficient than going to the expense and frustration of spending time in a waiting room, only to find once you get into the doctor's office that he or she is not open to the concept of natural therapies.

In such a letter you can ask for a short interview to discuss natural hormones and your philosophy concerning PMS, menopause, and perhaps even cancer. Every doctor's attitude is different. Some do not take kindly to their patients' relating to them information of which they were unaware. Then again, others, especially those who are not afraid of what their peers will think, may be quite receptive to natural options.

This is just the beginning of a portfolio outlining steps you can take to obtain NHRT through your medical doctor, who can call in a prescription (under your insurance coverage) to specialized pharmacies throughout the country. Appendix G provides a list of pharmacies that specialize in compounding natural hormones. Or, should you not have insurance, you can obtain the natural progesterone cream on your own directly from any of the distributors.

> All of us do better, physically, psychologically, and even physiologically, when we take an active, involved role in our own process of getting better—the exact opposite of "handing it over to the experts and hoping for the best."[20]
>
> *Stuart M. Berger, M.D.,*
> *What Your Doctor Didn't Learn in Medical School*

WAYS AND MEANS
OF HORMONE APPLICATION

Natural progesterone, estrogen, and even testosterone can be dispensed as tablets, creams, implants, injectable solutions, suppositories, or skin patches. Some of these products are to be administered by your medical doctor (be sure to request natural hormones only), while others you can obtain and use on your own. Many of the progesterone and estrogen creams, for example, come with written instructions on when, how, and where to use them. All of the items listed below are natural unless otherwise specified. For more information, call the source numbers listed in chapter 7.

USP PROGESTERONE VERSUS WILD YAM CREAMS

It is important to clarify that the progesterone cream used in Dr. John Lee's studies, and in most of the research cited, specifically contains USP progesterone, which is referred to as "natural" progesterone or simply "progesterone" on the labels. The major difference between this and the many wild yam creams available is in the laboratory processing, which converts the sapogenin molecule (most commonly from either the wild yam or the soybean) to a substance that duplicates the activity of the progesterone normally produced by the ovaries at the time of ovulation. It also acts as a precursor to other hormones as they are needed.[1]

Contrary to much of the skepticism regarding the wild yam products, many appear to be bioactive and also able to promote hormonal balance. In fact, many women have reported relief of their debilitating and draining symptoms of PMS, perimenopause, and menopause using products without this conversion. Some of these products containing wild yam or soy derivatives are known to provide results similar to those enjoyed by the Trobriand Islanders (whose diet consisted mainly of wild yam, other vegetables, and fish[2]), or the typical relatively asymptomatic experience of Asian women at the time of menopause, which is often attributed to their classic Oriental diet including a high amount of soy-based products, vegetables, and seafoods.[3]

TRANSDERMAL PROGESTERONE CREAMS

In determining how much transdermal progesterone cream to use, we need to be aware that optimal ovarian production of progesterone is somewhere in the range 15 to 30 mg per day from ovulation to menses. However, if a woman becomes pregnant, the placenta also begins a steadily increasing production of progesterone, reaching an upper limit of 300–400 mg per day during the third trimester. This is a wide margin, in contrast to the narrow thresholds of many of the body's other physiological ranges.

As progesterone only lasts in the body for six to eight hours, it's important to apply $\frac{1}{4}$ to $\frac{1}{2}$ teaspoon at least once in the morning and once in the evening. Progesterone cream can be massaged directly into the skin almost anywhere it's thin or soft—such as the wrists, inner arms, back of the hands, chest, breasts, lower abdomen, inner thighs, back, soles of the feet, face, and neck. It is preferable to alternate application among these various areas of the body to retain receptor sensitivity and to avoid wasteful oversaturation of any one area. The progesterone travels through the skin into the subdermal fat and then into the bloodstream. It is available either by prescription or in nonprescription form. Potency will vary, depending on your doctor's prescription and the individual manufacturer. Suggestions for specific uses follow.

Premenstrual Syndrome (PMS)/Perimenopause

Some instructions say that the amount needed by a cycling female will vary according to the degree of symptoms. If a woman is still having periods, progesterone works effectively when administered just prior to ovulation through just prior to menses. To determine when you are ovulating, please refer to the specific details in chapter 3. In more severe cases of PMS or perimenopause, start using the cream on approximately day 12, counting the first day of your last mentstrual period as day 1; and continue its use until day 26 or 27 (just before your next period begins). It is the sudden decrease in progesterone levels that triggers the menses a day or two later.

For minor symptoms, use the cream for fewer days, such as only ten days per month (count to day 16 from your last menstrual period and use ¼ to ½ teaspoon once or twice daily until day 26). This would provide the minimal amount and length of time necessary to begin buildup of adequate progesterone levels.

If you have cramping, apply to the abdomen as frequently as every half hour as needed for relief. If you suffer from menstrual migraines, you can apply the cream to your temples and to the back of your neck until the pain eases.

Should you become pregnant while using the cream, I recommend that you carefully read the documented information in chapter 3 concerning the use of progesterone during pregnancy. The benefits from this kind of hormonal support cannot be emphasized enough.

Menopause/Postmenopause/Posthysterectomy

Apply the cream daily for approximately three weeks out of every month. Abstaining for a week (or at least four days) each month helps maintain the body's ability to absorb it and reap maximum benefits. Normal use is approximately one jar or tube per month. If you do not notice a reduction of your symptoms within one to three monthly cycles of use, evaluate how often and how much cream you are using at each application. Many women, upon questioning, admit that they are inconsistent in their use of the cream, and many report dabbing it as if it were a rare perfume; sporadic or scant application can prevent the desired relief.

Calculating your approximate daily amount is easier to do when using a USP progesterone cream because the total active amount is listed in milligrams. Many jars or tubes contain 2 ounces, with 400 mg or more of progesterone per jar or tube. If you divide the 400 mg by the total number of days you use the cream each month (e.g., twenty to twenty-one days for menopausal women) and use one tube or jar each month, you will get approximately 20 mg or more per day. This daily amount can then be divided into two applications (three only if symptoms are severe), providing for a more continuous release of progesterone into the bloodstream.

Remember, osteoporosis is symptomless until the fracture threshold is reached, so if any bone density loss has been established, you'll need to continue a regular usage program in spite of the fact that you're no longer suffering from hot flashes, night sweats, and so on. We must not forget that for the treatment and reversal of osteoporosis, it is the USP-grade progesterone that to date has been clinically tested and evaluated.[4]

TABLETS OR CAPSULES: WOMEN ON A CYCLE/MENOPAUSE/POSTMENOPAUSE

With your doctor's guidance, follow the above instructions (as appropriate to your situation) for this delivery system also. Years ago, natural progesterone taken orally in pill form disintegrated in the stomach and never entered into the system. Now, certain pharmacies (see appendix G) have developed micronized progesterone in an oil base. The oil protects it from the acids in the stomach, and this process helps the progesterone make its way intact into the system for a better blood level.

You will notice when taking the capsule form of progesterone that the daily dosage is much higher (100 to 200 mg) than the cream form (20 to 24 mg). This higher amount compensates for the large percentage that's excreted by the liver. Its absorbability depends on many factors, such as the general health of a woman, how well her digestive system is functioning, and the health of her liver. As Dr. Lee says, "Oral doses of progesterone (even in micronized version) must be greater than transdermal doses to create the equivalent biologic effects."[5]

IMPLANTS (SYNTHETIC OR NATURAL)

This surgical procedure is done by your medical doctor. Once the hormone is implanted under the skin, women do not have to come back more than three times a year. However, if using the synthetic hormone, keep in mind what the authors of *Women on Menopause* say: "Once the implant is in place, you cannot get rid of it until the full six months is up, however badly you react to it."[6] Even after six months, some implants are reportedly difficult to remove.

A nineteen-year-old with a hormone implant came to my home in desperation because of the agonizing pain and pressure in her breasts. She claimed to no longer have any periods and was experiencing severe depression. The frightening physical symptoms and emotional instability that these toxic implants had brought on made her run for help in search of a natural alternative. But, unfortunately, the side effects from this drug will remain in her body for quite some time. The risks clearly do not seem to be worth any benefit she thought she was receiving.

It would take another book to tell of the horrendous consequences of having the synthetic drug implanted under the skin. Just a few obvious signs are excessive weight gain and extreme inflammation of the breast. My concern is for those women in similar situations who are not able to find supportive help. Hearing the horror stories of young girls on synthetic birth control implants would make anyone understand why we need to address this often overlooked issue.

PATCH (SYNTHETIC OR NATURAL)

This is a transdermal agent (that is, applied to the skin). It usually comes in an estrogen form prescribed by your medical doctor. With this method the estrogen is released through the skin in a consistent manner. The alcohol in the patch drives the hormone through the skin and into the blood vessels.

INJECTIONS (SYNTHETIC OR NATURAL)

Hormones are injected deep into the muscle. Some women are receiving progesterone this way, but the medical profession has not found this route as satisfactory or efficient as others. There is a question about the predictability of absorption rates. Also, in case you are considering the injections as a lifetime approach, they are just plain inconvenient.

SUPPOSITORIES (SYNTHETIC OR NATURAL)

Progesterone suppositories made with spermaceti wax are not considered very efficient by some medical doctors. Dr. Raymond Peat warns that "the bulk of the progesterone goes out of the solution very quickly, forming crystals which are essentially insoluble in body fluids. . . . Consumers are paying a high price for a minimal effect."[7]

SUBLINGUAL OIL

Dr. Richard Kunin reports that his patients experience relief from their PMS symptoms in less than fifteen minutes with a sublingual application of oil-based progesterone. Dr. Martin Milner also recommends an oil-based suspension for his pre-, peri-, and postmenopausal patients. Dr. Betty Kamen suggests use of sublingual progesterone for hot flashes and migraines as well as cramping at fifteen-minute intervals until symptoms disappear. This therapy can be used in conjunction with ¼ to ½ teaspoon of cream if symptoms are severe.[8]

CREAMS FOR VAGINAL ATROPHY

Pure progesterone, or estriol (possibly of risk to some women), is often locally applied to the vaginal wall, making it thicker, more elastic, and less susceptible to infection. Vaginal atrophy and the accompanying infections can cause severe pain during intercourse but can now be a condition of the past. You can also apply your progesterone or wild yam cream intravaginally at bedtime or as needed. (An example of how a bit of education might contribute to a happier and more relaxed life!)

The above application suggestions have been extracted from various educational materials and are based on the experiences of women in treating their premenstrual or menopausal problems. Of course, the dosage and method will vary according to individual needs, lifestyles, and preferences. The natural over-the-counter and prescription remedies can be obtained from any of the distributors or pharmacies listed in appendix G. I encourage you to ask questions and seek out supportive health-care providers who can guide you on your journey to optimal health.

Notes

1. Marcus Laux and Christine Conrad, *Natural Woman, Natural Menopause* (New York: HarperCollins, 1997), 74; and John R. Lee, *What Your Doctor May Not Tell You About Menopause* (New York: Warner Books, 1996).
2. Betty Kamen, *Hormone Replacement Therapy: Yes or No?* (Novato, CA: Nutrition Encounter, Inc., 1993), 216; *National Geographic*, October 1992.
3. H. Aldercreutz, "Dietary Phyto-estrogens and the Menopause in Japan," *Lancet*, Vol. 339, 1992, 1233.
4. John R. Lee, *What Your Doctor May Not Tell You About Menopause*.
5. Ibid, 266–267.
6. Anne Dickson and Nikki Henriques, *Women on Menopause* (Rochester, VT: Healing Arts Press, 1988), 59.
7. Raymond F. Peat, "Effectiveness of Progesterone Assimilation for the Relief of Premenstrual Symptoms" (educational brochure), Eugene, OR.
8. Kamen, 109-212.

SYNTHETIC COMPOUNDS: A SAMPLING BY CHEMICAL OR BRAND NAME

Below is just a handful, selected from the huge number of synthetically produced hormones and related compounds that are on the market today. Names continue to be added to the list periodically as women prove unable to tolerate the ones formerly prescribed. With the aid of William Boyd's *Textbook of Pathology*,[1] the *Physicians' Desk Reference* (PDR),[2] *Goodman and Gilman's The Pharmacological Basis of Therapeutics*,[3] and *Facts and Comparisons* ("the pharmacist's Bible"—updated monthly),[4] we have isolated some specific names that might be indicated on prescription labels. Following are some of the popular terms:

ESTROGENS

Estrone (Estrone Aqueous, Estronol, Theelin Aqueous, Kestrone, Estroject-2, Gynogen, Kestrin Aqueous, Wehgen)
Estradiol (Estraderm, Estrace)
Estradiol valerate (Delestrogen, Valergen 10, Valergen 40 Dioval, Duragen-20, Estra-L 20, Estra-L 40, Deladiol-40)
Conjugated estrogens (Premarin)
Esterified estrogens (Estratab, Menest)
Estropipate (Ogen, Ortho-Est, Estropipate)
Quinestrol (Estrovis)
Ethinyl estradiol (Estinyl)

DES (diethylstilbestrol)

Chlorotrianisene (Trace)

Estradiol cypionate (depGynogen, Depo-Estradiol Cypionate, Depogen, Dura-Estrin, Estra-D, Estro-Cyp, Estroject-L.A.)

Ortho Dienestol, DV, and many more.

Warnings and Adverse Reactions: "There is now evidence that estrogens increase the risk of carcinoma of the endometrium."[5] Other risks include excessive reproduction of normal cells on the inside lining of the uterus, breast cancer, gallbladder disease, and fluid retention, influencing the conditions of asthma, epilepsy, and cardiac or kidney dysfunction. Excess estrogen could bring on nausea, bloating, cervical discharge, polyp formation, skin discoloration, hypertension, migraine headache, breast tenderness, or edema. If contemplating surgery, discontinue use of estrogen at least four weeks prior to surgery to avoid the risk of blood-clotting complications.[6]

PROGESTINS

Medroxyprogesterone acetate (Provera, Amen, Curretab, Cycrin)

Hydroxyprogesterone caproate (Duralutin, Gesterol L.A., Hylutin, Hyprogest 250)

Norethindrone (Norlutin)

Norethindrone acetate (Norlutate, Norgestrel, Aygestin)

Megestrol acetate (Megace)

Micronor, Nor-Q.D., Ovrette, and many more.

Warnings and Adverse Reactions: (1) May cause insomnia, weakness, abdominal discomfort, flatulence, nausea, fever, vomiting, bleeding irregularities, erosion and abnormal secretions of the cervix, jaundice, severe itching, rash, acne, brown spots; also, risk of thrombophlebitis and thromboembolic disorders (blood clots in the lung, brain, or heart); cerebral hemorrhage and other cerebrovascular disorders; mental depression; impaired liver function; carcinoma of the breast. (2) Other risks include partial or complete loss of vision and retinal blood clots. Discontinue use if migraines, double vision, astigmatism, subluxated lens, apparent derangement of extraocular muscles, cataract, migraine, or edema should occur, or inflammation of the optic nerve at its point of entry into the eyeball. (3) Exposure to the progestins listed above that are prescribed during pregnancy can cause genital deformity in male and female fetuses (e.g., urethra abnormality). On the other hand, natural progesterone has been used successfully in premature labor to avoid spontaneous abortions and is of benefit to both mother and embryo (see chapter 3). (4) Careful observation should be paid to conditions brought on by fluid retention. Edema can affect epilepsy, migraine, asthma, and cardiac or renal dysfunction. Progestins in excess can cause weight gain, fatigue, abnormal menstrual flow, acne, hair growth or loss, depression, candida vaginitis.[7]

CONTRACEPTIVES (ESTROGEN/PROGESTIN)

Norgestrel, levonorgestrel, desogestrel, norgestimate, ethynodiol diacetate, norethindrone acetate, norethindrone, norethynodrel
Monophasic oral contraceptives (Genora, Norethin, Norinyl, Ortho-Novum, Ovcon, Demulen, Ovral, Nelova, Brevicon, Modicon, Ortho-Cyclen, Loestrin, Lo/Ovral, Desogen, Ortho-Cept, Levlen, Levora, Nordette)
Biphasic oral contraceptives (Jenest, Ortho-Novum)
Triphasic oral contraceptives (Tri-Norinyl, Tri-Levlen, Triphentasil, Ortho Tri-Cyclen)

Warnings: Clots in blood vessels to the heart and lungs may cause strokes or heart failure, retinal thrombosis, and optic neuritis. Other risks: gallbladder disease; insulin resistance, breast presssure or pain, cervical cancer, depression, abdominal pain and tenderness, elevated blood pressure, decreased glucose tolerance, excessive calcium in the blood. In chapter 4 we learned that natural micronized progesterone builds bone mineral density (BMD). On the other hand, "Medroxyprogesterone may be considered among the risk factors for development of osteoporosis."[8] Other side effects: photosensitization, abnormal or excessive uterine bleeding, fluid retention and related conditions of asthma, epilepsy, migraine, and heart, kidney, and liver disorders (benign or malignant tumors which may rupture or hemorrhage and cause severe abdominal pain, shock, or death).[9] Likewise, if one already suffers from hypertension, obesity, or diabetes, the prescription of contraceptives may cause sickness or even death.[10]

PROGESTIN/ESTROGEN COMBINATIONS

We find over forty different names, beginning with Brevicon and ending with Triphasit-28, listed in the latest *Facts and Comparisons*. For more details, call the number below or look up the most recent *Physicians' Desk Reference* (PDR) in your local library.

The *Warnings and Risks,* according to *Facts and Comparisons,* are much the same for this section. For updated information on new hormones or current prescriptions, call (800) 223-0554.

Notes

1. William Boyd, *Textbook of Pathology (8th ed.),* (Philadelphia, Lea & Febiger, 1970), 18.
2. *Physicians' Desk Reference* (Montvale, NJ: Medical Economics Data Production Company, 1995).
3. Alfred Gilman and Louis Goodman, *Goodman & Gilman's Pharmacological Basis of Therapeutics* (6th ed.) (New York: Macmillan, 1980), 1420–1438.
4. *Facts and Comparisons* (St. Louis, MO: Facts and Comparisons, Inc.), 96–108 (looseleaf booklet published monthly: 111 West Port Plaza, Suite 400, St. Louis, MO, 63146).
5. Ibid.
6. Ibid.
7. Ibid.
8. Ibid.
9. Ibid.
10. Ibid.

NATURAL FORMULAS
FOR INFANTS

In chapter 6 we discussed soy allergies and particularly why people seem more sensitive to highly processed soy products. Sally Fallon, M.A., a member of the Price Pottenger Nutrition Foundation Advisory Board, and Mary Enig, Ph.D., nutritionist and expert in the field of lipid chemistry, provide some excellent recipes for baby formula. More information on their work can be found in their book *Nourishing Traditions: The Cookbook That Challenges Politically Correct Nutrition and the Diet Dictocrats,* as well as in their well-researched articles published in *Health Freedom News.*[1]

When making baby formula, the authors encourage the use of whole foods instead of isolated soy protein. And according to the Community Nutrition Institute, "Scientists assert that neonatal infants are particularly vulnerable to estrogens and that insufficient research on the long-term health effects of phytoestrogens warrants a ban on the nonprescription sale of soy formula."[2] The following are some nourishing baby formulas. It's interesting to note that gelatin added to cow's milk not only emulsifies the fat, but it balances the casein (milk protein). This improves the absorption of the fat and also its digestibility. Studies show that "milk containing gelatin is more rapidly and completely digested in the infants" (N. R. Gotthoffer, *Gelatin in Nutrition and Medicine).*[3]

Soy-Free Infant Formula with Milk[4]

2 cups (16 oz) raw organic milk or
  cultured milk, not homogenized
¼ cup whey
4 T. lactose
1 tsp. cod liver oil
1 tsp. unrefined sunflower oil
1 tsp. extra virgin olive oil
2 tsp. coconut oil
2 tsp. brewers yeast
2 tsp. gelatin
1¾ cup filtered water
1 100 mg tablet vitamin C, crushed
(Never heat formula in a microwave oven!)

Soy-Free, Milk-Free Infant Formula[5]

3½ cups homemade broth (beef, lamb, chicken—hormone and
  antibiotic free—or fish)
2 ounces organic liver (liquify)
5 T. lactose
¼ cup whey
3 T. coconut oil
1 tsp. cod liver oil
1 tsp. unrefined sunflower oil
2 tsp. extra virgin olive oil
1 100 mg tablet vitamin C, crushed

Notes

1. Sally W. Fallon and Mary G. Enig, "Soy Products for Dairy Products? Not So Fast. . . ." *Health Freedom News*, Vol. 14, No. 5, September 1995.
2. Sally Fallon, Pat Connolly, and Mary G. Enig, *Nourishing Traditions: The Cookbook That Challenges Politically Correct Nutrition and the Diet Dictocrats* (San Diego: ProMotion Publishing, 1995), 564.
3. Ibid, 561.
4. Fallon and Enig, "Soy Products for Dairy Products?"
5. Ibid.

RESOURCES FOR CANCER PATIENTS

MYTHS OF THE MAMMOGRAM

Again and again we are asked to have yearly mammograms, but you might want to think twice the next time you see TV ads leading you to believe that regular screening may "add years to your life." Who would think that it might actually take years away from your life? The problem is that we can't observe any major side effects immediately after having a mammogram. However, these may begin to arise as time goes on. Thus, it is extremely important for women to educate themselves so that they can face the barrage of misleading information on the subject.

But this can be a confusing task. The leading cancer organizations have flip-flopped repeatedly over the years on their advice to women in various age groups. Despite what their most recent recommendation might be, note that back in 1976 the *New England Journal of Medicine* reported that the National Cancer Institute (NCI) and the American Cancer Society had "terminated the routine use of X-ray mammography for women under the age of fifty because of its detrimental effects."[1] And in 1994, the National Cancer Institute reported "no difference in the fatality rate [among women in their forties] between those whose breast cancer was detected by mammogram or those diagnosed by touch or palpation alone...therefore, in the NCI's judgment, there's no advantage in subjecting these women to regular screening."[2]

William C. Bryce, M.D., founding director of the Well Breast Foundation in California, also suggests the alternatives of thermography and diaphanoscopy, or transillumination of the breast, in his article "The Truth

About Breast Cancer" *(Health Freedom News)*. Says Dr. Bryce, "Until the question of lifetime risk due to radiation can be resolved, mammography . . . should not be used in regular routine screening procedures. . . . A few years ago the National Cancer Institute stated that for every fifteen cancers diagnosed by mammography in women under thirty-five, seventy-five were caused [by mammograms]."[3]

The preventive measures proposed by medical doctors can be costly in more ways than one. Marcus Laux, N.D., reminds us that radiation damage to the cells of the body is cumulative over time. "There is increasing evidence," he says, "that the ionizing radiation used in mammograms is not only harmful, but can even cause the very cancer it's supposed to detect! Many pre-eminent cancer researchers, like Dr. John Bailar III, former editor of the *Journal of the National Cancer Institute,* predict that routine mammogram screening over a period of 10 to 15 years may induce breast cancer."[4] Dr. Laux reports that animal studies have shown that pressure on and manipulation of tumors (as might occur during mammography) causes an 80 percent increased chance that they will grow and metastasize![5]

Dr. Maureen Roberts wrote a powerful critique of mammographic screening prior to her own death from breast cancer. In an article ("Breast Screening: Time for a Rethink?") published in the *British Medical Journal,* she stated her reservations—that mammography is an unfit screening test in that it's "technologically difficult to perform, the pictures are difficult to interpret, it has a high false positive rate, and we don't know how often to carry it out. We can no longer ignore the possibility that screening may not reduce mortality in women of any age. . . . If screening does little or no good, could it possibly be doing any harm?"[6]

In a similar vein, *The Lancet* reported in 1985 that over 280,000 women were recruited in order to evaluate the potential risk of breast cancer by mammography. The conclusion was: "Even if you catch it early, with orthodox medicines and chemotherapy, you will not survive any better than if you catch it late,"[7] given the questionable outcome of conventional cancer therapy.

In this controlled trial for women below fifty, none was advised about the potential risks. Despite warnings by a committee of the United States' National Academy of Sciences, the women were exposed to doses that could possibly cause more cancer in the long run than could be prevented by the program.[8] In a ten-year follow-up of the mammography study, the women who refused screening (35 percent) had a lower incidence and mortality from breast cancer than either the mammography group or the control group.[9]

We can look to the discussion in *Healthy Healing* for a good answer to the question, "Should I or should I not have a mammogram?" Dr. Rector-Page warns that, "Although mammograms have improved in the last 20 years both in clarity and amount of dosage, we still hear enough horror stories about swift fibroid onset to recommend that mammograms should not be done routinely or without suspected cause." She also notes that mammograms and low-dose X rays may result in iodine depletion and thyroid problems.[10] She continues with the most prudent statement I have heard yet on the subject: "While early detection can mean less radical medical intervention, prevention through immune enhancement and a healthy lifestyle should be the primary goal—not early detection."[11]

One alternative to the mammogram is the sonogram, which uses sound waves rather than electromagnetic waves and is, according to author Dee Ito, particularly accurate "in determining the presence of invasive cancers [and] cystic breasts."[12] Nonetheless, further studies are needed as there are preliminary indications that even sonography produces some cellular changes in the tissues.

There is so much information about the potential risks resulting from routine mammograms that it is essential for all women to look into the matter for themselves and make informed decisions. For more information on radiation levels in mammography, important questions to ask prior to scheduling a mammogram, and on the funding and politics behind cancer treatments, I urge you to read Virginia M. Soffa's *The Journey Beyond Breast Cancer* as well as other related books in the suggested reading list that follows.

USEFUL READING

Betrayers of the Truth: Fraud and Deceit in the Halls of Science by William Broad & Nicholas Wade, New York: Simon & Schuster, Inc., 1983.

Cancer Prevention and Nutritional Therapies by Richard A. Passwater, Ph.D., New Canaan, CT: Keats Publishing, Inc., 1994.

Cancer Therapy: The Independent Consumer's Guide to Non-toxic Treatment & Prevention by Ralph W. Moss, Ph.D., New York: Equinox Press, 1992.

Choices in Healing by Michael Lerner, Cambridge, MA: The MIT Press, 1994.

The Healing of Cancer by Barry Lynes, Ontario, Canada: Marcus Books, 1990.

How to Get Well by Paavo Airola, Phoenix: Health Plus Publishing, 1985.

How to Prevent Breast Cancer by Ross Pelton. New York: Simon & Schuster, 1995.

The Journey Beyond Breast Cancer by Virginia M. Soffa, Rochester, VT: Healing Arts Press, 1994.

Love, Medicine & Miracles by Bernie S. Siegel, M.D., New York: Harper & Row, 1990.

Now That You Have Cancer by Drs. Robert W. Bradford and Michael L. Culbert, Chula Vista, CA: Bradford Foundation, 1992.

The Persecution and Trial of Gaston Naessens by Christopher Bird, Tiburon, CA: Krammer Inc., 1991.

The Scientific Validation of Herbal Medicine by Daniel B. Mowrey, Ph.D., New Canaan, CT: Keats Publishing, Inc., 1990.

Sharks Don't Get Cancer by William I. Lane and Linda Comac, New York: Avery Publishing Group, Inc., 1992.

Third Opinion by John M. Fink, Garden City Park, New York: Avery Publishing Group, Inc., 1988.

The books listed above provide information on alternatives to conventional cancer treatment. I strongly recommend you also contact the following organizations:

ORGANIZATIONS

Foundation for Advancement in Cancer Therapy (F.A.C.T.)
Box 1242 Old Chelsea Station
New York, NY 10113
(212) 741-2790

The International Association of Cancer Victors and Friends, Inc.
7740 West Manchester Avenue, Suite 110
Playa del Rey, CA 90293
(213) 822-5032 or (213) 822-5132.

CANHELP, Inc.
3111 Paradise Bay Road
Port Ludlow, WA 98365-9771
(206) 437-2291

People Against Cancer
Frank D. Wiewel, Executive Director
P.O. Box 10
Otho, IA 50569
(515) 972-4444

Cancer Control Society
2043 North Berendo Street
Los Angeles, CA 90027
(213) 663-7801

Bradford Research Institute
1180 Walnut Ave.
Chula Vista, CA 91911

Center for Advancement in Cancer Education
Wynnewood, PA 19096
(610) 642-4810

The University of Natural Healing, Inc.
P.O. Box 8113
Charlottesville, VA 22906

INSURANCE COVERAGE

For information on insurance for cancer treatment, write to the Association of Community Cancer Centers, 11600 Nebel Street, Suite 201, Rockville, MD 20852. For a list of the naturopathic physicians in your area, write to the American Association of Naturopathic Physicians (A.A.N.P.), 2366 East Lake Avenue East, Suite 322, Seattle, WA 98102, or call (206) 328-8510.

Also, if you prefer to see a doctor who does not prescribe synthetic medication, but you find your insurance does not cover his or her services (naturopathy, homeopathy, chiropractic, acupuncture, bodywork and massage, biofeedback, Ayurvedic medicine, herbal remedies, nutritional counseling, Oriental medicine, etc.), you might want to contact Alliance for Alternatives in Healthcare, Inc. at (805) 494-7818, P.O. Box 6279, Thousand Oaks, CA 91359-6279.

A BRIEF INTRODUCTION TO NATURAL CANCER THERAPY

Plant-Based Dietary Elements

> Fresh, raw fruits and vegetables
> Whole grains (including brown rice)
> Vegetable protein; raw nuts and seeds
> Raw (bitter) almonds (containing laetrile)

Olive and coconut oils

Soluble fiber

Fasting with fresh fruit juices in A.M. and vegetable juices in P.M.
(lemon, apple, orange; carrot, beet, spinach, chard, cabbage, kale,
parsley, asparagus, tomato, etc.)

Onions

Aged garlic extract

Yams

Mushrooms (reishi, shiitake, etc.); Krestin (Japanese mushroom
extract)

Aloe vera

Sea vegetables, including kelp (laminaria)

Algae (chlorella, spirulina, dunaliella bardawil, sargassum,
kjellmanianum)

Green barley

Wheatgrass

Chlorophyll

Lactobacilli (e.g., acidophilus, bifidis, and bulgaricus, in yogurt and
other cultured milk products)

Amino acids (including l-carnitine, l-cysteine, l-glutathione, l-
methionine, etc.)

Vitamins, Minerals, Antioxidants, Enzymes

Vitamins A, E, C (with bioflavanoids), D, K, B-complex; beta
carotene and other carotenoids

Calcium

Magnesium

Zinc

Tellurium

Selenium

Pycnogenol

Germanium

Curcumin

Coenzyme Q_{10}

S.O.D.

Proteolytic enzymes

L-Asparaginase

Bromelain

Herbs, Teas, Spices

Hoxsey herbs
Chinese medicine
Astragalus
Echinacea
Black radish
Chaparral
Dandelion
Silymarin (milk thistle)
Pau d'arco
Red clover
Rhubarb
Slippery elm
Suma
Iscador
Actinada
Kermesbeurro
Viva natural extract
Jason Winters tea
Green tea
Essiac tea
Turmeric
Cumin seeds
Basil leaves
Black pepper
Poppy seeds
Cinnamon
Asafoetida
Drumstick leaves
Kandathipili
Manathakkahli leaves
Neem flowers
Ponnakanni

Miscellaneous

Bovine cartilage
Camphor derivative (714-X)
Castor oil packs

Colon cleansing
Detoxification therapy
DHEA
EDTA chelation therapy
Electromagnetic treatment (Bjorn Nordenstrom)
Intravenous amino acids/vitamins
Intravenous vitamin C
Hyperthermia
Homeopathic liver drainer
Live cell therapy
Liver flushes
Oxygen therapy
pH balancing
Phototherapy
Shark cartilage suramin
Thioproline
Tumostereone

This list is only a beginning. It should be viewed in the context of the whole-body approach to creating *host resistance* to disease. There are many more counter-agents to cancer, as well as vaccines and drugs that contain synthetic components and therefore are not listed here. Please consult the references for more information, and study all the pros and cons that may be involved.

Notes

1. Lorraine Day, *Cancer Doesn't Scare Me Anymore!* (videotape) (Rancho Mirage, CA: Rockford Press, 1994).
2. Stanley Englebardt, "Straight Talk About Mammograms," *Reader's Digest*, November 1994.
3. William C. Bryce, "The Truth about Breast Cancer," *Health Freedom News*, May 1993, 11.
4. Marcus Laux, "Why I Don't Think You Should Trust Mammograms," *Naturally Well*, Vol. 3, No. 5, 1996, 2.
5. Ibid; *Archives of Internal Medicine*, Vol. 56, No. 2, 1996, 209–213; W. W. Fletcher, "Why Question Screening Mammography for Women in Their Forties?" *Radiol. Clin. North America*, Vol. 33, No. 6, 1995, 1259–1271; Karla Kerlikowski, "Efficacy of Screening Mammography: A Meta-analysis," *Journal of the American Medical Association*, Vol. 273, No. 2, 1995; K.

Lockwood, S. Moesgard, and K. Folders, "Partial and Complete Regression of Breast Cancer in Patients in Relation to Dosage of Coenzyme Q10," *Biochem. Biophys. Res. Commun.*, Vol. 199, No. 3, 1994, 1504–1508.

6. Maureen Roberts, "Breast Screening: Time for a Rethink?" *British Medical Journal*, Vol. 299, 1989, 1153–1155.

7. P. Skrabanek, *The Lancet*, August 10, 1985, 316–320.

8. Ibid.

9. Samuel S. Epstein, *The Washington Post*, March 10, 1992; *Cancer Watch*, People Against Cancer, Winter/Summer 1993.

10. Linda Rector-Paige, *Healthy Healing: An Alternative Reference* (Soquel, CA: Healthy Healing Publications, 1992), 166, 329.

11. Ibid, 167.

12. Dee Ito, *Without Estrogen: Natural Remedies for Menopause and Beyond* (New York: Carol Southern Books, 1994), 14.

RESOURCES FOR PREVENTIVE MEDICINE

USEFUL READING*

Alternative Medicine by The Burton Goldberg Group. Puyallup, WA: Future Medicine Publishing, Inc., 1993.

Aspartame (NutraSweet): Is It Safe? by H. J. Roberts. Philadelphia: Charles Press, 1990.

The Bitter Truth About Artificial Sweeteners by Dennis W. Remington and Barbara W. Higa. Provo, UT: Vitality House International, 1987.

The Chelation Way by Dr. Walker Morton. New York: Avery Publishing Group, 1990.

Choices in Healing by Michael Lerner. Cambridge, MA: MIT Press, 1994.

Coenzyme Q-10 by William H. Lee, R.Ph., Ph.D. New Canaan, CT: Keats Publishing, Inc., 1987.

Food Enzymes by Humbart Santillo, B.S., M.H. Prescott, AZ: Holm Press, 1991.

Fresh Vegetable and Fruit Juices by N.W. Walker, D.Sc. Prescott, AZ: Norwalk Press, 1978.

Garlic: Nature's Original Remedy by Stephen Fulder and John Blackwood. Rochester, VT: Healing Arts Press, 1991.

Garlic for Health by Benjamin Lau, M.D., Ph.D. Brushton, NY: Teach Services, 1988.

Green Barley Essence, The Ideal Fast Food by Yoshihide Hagiwara, M.D. New Canaan, CT: Keats Publishing, Inc., 1986.

---

*For books on NHRT, see appendix G.

Healthy Healing: A Guide to Self-Healing for Everyone by Linda G. Rector-Page, N.D., Ph.D. Soquel, CA: Healthy Healing Publications, 1997.

Healing Nutrients by Patrick Quillin, Ph.D., R.D. New York: Vintage Books, 1986.

How to Get Well by Paavo Airola, Ph.D. Phoenix, AZ: Health Plus Publishers, 1980.

Jane Brody's Nutrition Book by Jane Brody. New York: Bantam Books, 1982.

The Nutrition Desk Reference by Robert H. Garrison, Jr., M.A.R., Ph.D., and Elizabeth Somer, M.A., R.D. New Canaan, CT: Keats Publishing, Inc., 1990.

Prescription For Nutritional Healing by James F. Balch and Phyllis A. Balch, C.N.C. Garden City Park, NY: Avery Publishing Group Inc., 1990.

A Reference Guide for Essential Oils, compiled by Pat Leathan and Connie Higley. Topeka, KS: Abundant Health, 1996.

Seasalt's Hidden Powers by Jacques de Langre, Ph.D. Magalia, CA: Happiness Press, 1993.

Sunlight by Zane R. Kime, M.D., M.S. Penryn, CA: World Health Publications, 1980.

An Introduction to Young Living Essential Oils and Aromatherapy by D. Gary Young, N.D. Scottsdale, AZ: Essential Press, 1996.

Your Body's Many Cries for Water by F. Batmanghelidj, M.D. Falls Church, VA: Global Health Solutions, Inc., 1992.

What Your Doctor Didn't Learn in Medical School by Stuart M. Berger, M.D. New York: William Morrow and Company, Inc., 1988.

Homeopathic Medicine

Everybody's Guide to Homeopathic Medicines by S. Cummings, M.D., and D. Ullman, M.P.H. J. P. Tarcher, Inc., 1980.

Everyday Miracles by Linda Johnston, M.D. Christine Kent Agency, 1991.

The Family Health Guide to Homeopathy by Barry Rose, M.D. Celestial Arts, 1993.

Homeopathic Medicine for Pregnancy and Childbirth by Richard Moskowitz, M.D. North Atlantic Books, 1992

ACCESS TO THE INTERNET

For the doctor who says, "Sorry, I just can't provide these natural hormones; we don't have enough information or studies in this area," you might want to hook up to the worldwide computer search network. You or your doctor can order and receive abstracts and publications on any medical subject— which would include some of the natural alternatives referred to in this book. For the latest information about natural progesterone and related topics, the day has arrived when you can find answers right on your computer screen. For instance, to find data on about progesterone, use your web browser to search for the term "natural progesterone." Be sure to include the word *natural* so you won't be inundated with information on the synthetic estrogen and progestin products. As of this writing, we obtained more than two hundred sites relating to natural progesterone. An example of one of the many sources on the Internet, where experts are available to answer questions online, is Debbie Moskowitz, N.D., the staff doctor at Transitions for Health. Her home page address (at the time of this writing and subject to change) is http://www.progest.com. Questions and replies are sent via confidential e-mail. If you feel more comfortable talking on the telephone, you may call (800) 888-6814.

Many women are finding this means of communication quite gratifying as they find friends to talk to about health or nutritional problems similar to their own. Thanks to the electronic bulletin boards, women are hearing from other women who currently use progesterone. So today, in an instant, we can find out that we're not alone and that our situation is not unique. Often we can learn from the shared experiences of others. Of course, we also need to be aware of the misinformation and outright scams to be found online since many web sites represent a particular business.

INFORMATION FOR YOUR PHYSICIAN

(Your Name)
(address)
(phone number)

Dear Doctor,

I am interested in finding a physician who can guide me in the use of natural hormone supplementation. Please read the attached information, which has been summarized from various medical reports. I would like to know whether you are willing to work with women who do not wish to use synthetic hormones, and whether you would prescribe natural progesterone and estriol if needed. If so, please have your nurse or receptionist contact me for an appointment.

Sincerely yours,

Enclosure

(The next several pages and the references at the end of this appendix can accompany this letter.)

Progesterone and Osteoporosis

Medical Hypotheses states that "progesterone and not estrogen is the missing factor . . . effective in reversing osteoporosis. . . . The presence or absence of estrogen supplements [in subjects studied] had no discernible effect on osteoporosis benefits." The journal also claims that the use of natural progesterone is not only safer but less expensive than using Provera (medroxy-progesterone) and that "progesterone deficiency rather than estrogen deficiency is a major factor in the pathogenesis of menopausal osteoporosis."[1] We need to be aware of the frightening accounts and adverse effects of Provera and other synthetic progestins (nervousness, depression, insomnia, immune and circulatory disorders, and much more).

Progesterone works to stimulate bone production, even when estrogen activity is low or absent. Because progesterone appears to work on the osteoblasts (bone-building cells) to increase bone formation, it would complement estrogen's action of decreasing bone resorption, as stated by Dr. J. C. Prior, who explains further that progesterone fastens to receptors on the osteoblasts and "increases the rate of bone remodeling."[2] Estrogen helps to slow bone loss, but progesterone is proactive through its stimulatory effect on the osteoblasts and thus directly encourages bone build-up. Synthetic progestins diminish the supply of natural progesterone, which further accelerates the osteoporotic changes within the bone.

Vitamin Therapy to Counter Estrogen Dominance

Carlton Fredericks, Ph.D, suggests that the use of vitamins A and E cream for atrophic vaginitis is much safer than the synthetic estrogen creams so frequently prescribed by many medical doctors. He says that "both [Cynonal and Premarin] have side effects and may also irritate you."[3] Of much greater concern, however, is the fact that Premarin consists of estrone and estradiol, which have been shown to be carcinogenic.[4] Fortunately, there are companies who make the natural progesterone cream and estriol creams that work wonders for vaginal dryness and many other conditions.

Dr. Fredericks, in his book *Guide to Women's Nutrition*, states that "vitamin B complex cuts down excess activity of estrogen, whether prescribed for menopause or birth control or produced by the woman's own ovaries." B complex, he emphasizes, "helps to counteract the effects of high levels of estrogen which has not been detoxified by the liver."[5]

Notes

1. John R. Lee, "Is Natural Progesterone the Missing Link in Osteoporosis Prevention and Treatment?" *Medical Hypotheses*, 1991, 35, 316, 318.
2. J. C. Prior, Y. Vigna, and N. Alojada, "Progesterone and the Prevention of Osteoporosis," *Canadian Journal of OB/GYN & Women's Health Care*, Vol. 3, No. 4, 1991, 181.
3. Carlton Fredericks, *Guide to Women's Nutrition* (New York: Putnam, 1989).
4. Follingstad, "Estriol."
5. Fredericks, *Guide.*

SOURCES OF NATURAL PROGESTERONE

T he natural progesterone in tablet or cream form, though approved by the FDA, is not available in most neighborhood drugstores. However, botanical hormones are accessible in various forms at dispensaries and health food stores around the country, some of which are listed on the following pages. As more and more physicians begin to understand their patients' needs for better assimilated hormones, additional pharmacies and stores will be encouraged to make these natural products available. The shift in consumer attitudes that is taking place has been noticed by these natural compounding pharmacies and their pharmacists, who report that women want natural products and "doctors are being forced to pay attention because women are bringing them the solution [to negative side effects]. They are demanding better treatment."[1]

Some of the larger firms listed here can also provide a variety of services that are usually not available in our neighborhood stores. For instance, regardless of which part of the country you live in, Women's International Pharmacy can (1) furnish you with at least a partial list of medical doctors located in your area who prescribe the natural hormones, (2) supply you with a list of local chiropractors or naturopaths who have ordered the nonprescription hormones (creams, gels, sublingual oils, etc.) for their patients' use, and (3) even file your insurance forms for you.

Not all of the druggists listed carry the nonprescriptive transdermal progesterone cream (labeled under a variety of different names). Nonetheless, these can be ordered through distributors and manufacturers, several of which are listed here. Increasingly, health food stores are also carrying this natural progesterone cream, which they in turn obtain from the distributor or manufacturer. Be sure to encourage your local health store manager to stock these products.

A few of the pharmacies listed here also offer patient consultation and education. But what is most important, they *all* carry oral micronized progesterone, which comes in tablet and capsule form (dosage ranging from 1 to 6 percent, 100 to 400 mg, depending on what your doctor prescribes) and is compounded according to the physician's prescription by a licensed pharmacist using FDA-approved ingredients. All these pharmacies carry estriol as well. Some physicians prescribe simply the estriol along with progesterone, while others prescribe a "tri-estrogen" (80 percent estriol, 10 percent estradiol, 10 percent estrone) with progesterone. With many diverse philosophies and differing patient requirements, prescriptions can be customized in various ways.

If you have personal questions concerning your individual health needs or want more information about ordering natural hormones in a variety of creams, gels, capsules, patches, sublingual oils, sublingual orange-flavored tablets, sprays, or suppositories, contact any of the following pharmacies, producers, or distributors. After you have discussed your requirements with them, it will be easier for you to decide what step you or your doctor should take next.

If your doctor says he can't get natural progesterone, you might want to tell him that according to the *New York Times* (November 18, 1994), "The Upjohn Company in Kalamazoo, Michigan derives the finely particled progesterone from soybeans and sells it to pharmacies as a bulk powder. Pharmacies can then formulate it into capsules, tablets, suspensions or suppositories in various dosages according to a physician's prescription."[2] Your doctor can also obtain micronized progesterone from Schering-Plough Corporation in Madison, New Jersey, by requesting the soybean derived hormone manufactured without the estrogen.

Today most USP progesterone is, in fact, extracted from soy. You will recall that neither USP nor human progesterone is present in either of the major plant sources (soybean or wild yam). Yams contain the sterol diosgenin, whereas soybeans contain the sterol stigmasterol—both of which have progesterone-like effects.[3] The substance sold as USP progesterone is produced in the lab by hydrolyzing extracts of soy or yam and converting saponins into sapogenins, two of which, sarsasapogenin (soy) and diosgenin (yam), provide the majority of derivation of natural progesterone produced for medical purposes. Actually, with the phytogenins from plants it is possible to manufacture not only progesterone but other hormones such as testosterone and estrogen.

However, nonprescriptive progesterone creams bought at health stores or from the manufacturers and distributors listed below are derived from the wild yam. The diosgenin from the wild yam is a precursor of progesterone. In fact, diosgenin is quite abundant in certain species of *Dioscorea*.

While diosgenin may have some progestogenic or even phytoestrogenic action, the effect varies from one person to another.[5] Some doctors say that the human body cannot convert wild yam or diosgenin to hormones and that conversion to progesterone must take place in a laboratory. It is possible, however, that some women's bodies are better able to utilize plant-derived compounds than others. It is also important to remember that while the mechanism of phytogenic activity may not be clearly understood at this time, botanical supplementation continues to gain support among women everywhere because it works for them.

You may hear charges that the use of wild yam to produce natural progesterone is threatening to make it an endangered species. I saw one such article just before going to press. Pharmacologist James Jamieson, however, assures me that this is definitely not the case. The wild yam is very common and grows all over the world in a wide variety of natural environments—Central America, Peru, China, Germany, and Afghanistan, to name a few. While it is possible that the plant may become scarce in particular locations, wildcrafting is very unlikely to have a serious impact on its worldwide distribution. In addition, these big, spongy tubers are easy to grow and already cultivated commercially. Since we all need to be aware of the effect our personal decisions have on the environment, this information about the wild yam is reassuring.

My life has been greatly simplified now that I know where to call for my prescriptive or nonprescriptive hormone replacement. Furthermore, I waste no more time driving to and from the drugstore and waiting in line. You, too, will find life less stressful when your products arrive by UPS or the U.S. mail at your front door. What could be easier than having your natural hormone therapy delivered direct to your home? Give yourself a break—you deserve this service!

Pharmacies

Apothecure, Inc.
13720 Midway Road
Dallas, TX 75244
(800) 969-6601

Artesia Pharmacy
18550 South Pioneer Boulevard
Artesia, CA 90701
(800) 851-7900

Bellgrove Pharmacy
1535 116th Avenue NE
Bellevue, WA 98004
(800) 446-2123

Belmar Pharmacy
12860 West Cedar Drive, Suite 210
Lakewood, CO 80228
(800) 525-9473

California Pharmacy and Compounding Center
307 Placentia Avenue, #0102
Newport Beach, CA 92663
(800) 575-7776

Clark's Pharmacy
15615 Bel-Red Road
Bellevue, WA 98008
(800) 480-DHEA

College Pharmacy
833 North Tejon Street
Colorado Springs, CO 80903
(800) 888-9358

Homelink Pharmacy
2650 Elm Avenue, Suite 104
Long Beach, CA 90806
(800) 272-4767

Hopewell Pharmacy
1 West Broad Street
Hopewell, NJ 08525
(800) 792-6670

Madison Pharmacy Associates and
  Bajamar Women's Healthcare
429 Gammon Place
P.O. Box 259641
Madison, WI 53791-9786
(800) 558-7046

Medical Center Pharmacy
10721 Main Street
Fairfax, VA 22030
(800) 723-9160

Miller Compounding Pharmacy
431 Commerce Park Drive,
  Suite 106
Marietta, GA 30060
(800) 547-1399

Pavilion Pharmacy
3193 Howell Mill Road, Suite
  122A
Atlanta, GA 30327
(404) 350-5780

Professional Arts Pharmacy
1101 North Rolling Road
Baltimore, MD 21228
(800) 832-9285

Ray's Pharmacy, Inc.
400 South Main Street
Manfield, TX 76063
(800) 255-7153

Snyder-Mark Drugs Roselle
225 East Irving Park
Roselle, IL 60172
(800) PROGEST/776-4378

Trumarx Drugs
501 Gordon Avenue
Thomasville, GA 31792
(800) 552-9997

University Compounding
    Pharmacy
550 Washington Street
San Diego, CA 92103
(800) 985-8065

Wellness Health &
    Pharmaceuticals
2800 South 18th Street
Birmingham, AL 35209
(800) 227-2627

The Women's International
    Pharmacy
5708 Monona Drive
P.O. Box 6468
Madison, WI 53716
(800) 279-5708

Manufacturers and Distributors

AIM
3904 East Flamingo
Nampa, ID 83687-3100
(800) 456-2462

Alvin Last Inc.
19 Babcock Place
Yonkers, NY 10071
(800) 527-8123

Angel Care U.S.A.
3666 North Peachtree Road,
    Suite 300
Atlanta, GA 30341
(800) 854-3895

Beyond A Century
HC 76 Box 200
Greenville, ME 04441
(800) 777-1324

Bio-Nutritional Formulas
106 East Jericho Turnpike
Mineola, NY 11501
(800) 950-8484

Dixie PMS and Menopause
    Center
2161 Newmarket Parkway, Suite 222
Marietta, GA 30067
(800) PMS-9232

Emerson Ecologics, Inc.
18 Lomar Park
Pepperell, MA 01463
(800) 654-4432

HealthWatchers System®
13402 North Scottsdale Road,
    Suite 150
Scottsdale, AZ 85254-4056
(800) 321-6917

International Health
8704 East Mulberry Street
Scottsdale, AZ 85251-5023
(800) 481-9987

Karuna Corporation
42 Digital Drive, Suite 7
Novato, CA 94949
(800) 826-7225

Kenogen
P.O. Box 5764
Eugene, OR 97405
(541) 345-9855

Nature's Nutrition, Inc.
4040 Lake Griffin Road
Lady Lake, FL 32159
(800) 242-1115

New Life Nutriceuticals
P.O. Box 996
Boca Raton, FL 33429
(800) 282-7216

Nutraceutics Corporation
600 Fairway Drive, Suite 105
Deerfield Beach, FL 33441
(800) 851-7007

NutriSupplies, Inc.
7901 79th Way
West Palm Beach, FL 33407
(800) 906-8874

Phillips Nutritionals
27071 Cabot Road, #121
Laguna Hills, CA 92653
(800) 514-5115

Professional & Technical Services
5200 S.W. Macadam Avenue,
    Suite 420
Portland, OR 97201
(800) 888-6814

Sedna Specialty Health Products
P.O. Box 347
Hannibal, MO 63401
(800) 223-0858

Young Again
43 Randolph Road, #125
Silver Spring, MD 20904
(301) 622-1073

For a list of more compounding pharmacies nationwide or closer to you, call Professionals and Patients for Customized Care at (713) 933-8400.

As of this writing, I find that more manufacturers and distributors could be added daily to this list. Because of the popularity and diversity of the health benefits of phytohormones, it is virtually impossible to keep up with all the new companies marketing such products.

There has been a great deal of confusion pertaining to the progesterone content of various manufacturers' transdermal creams. The bioavailability of the progesterone in such products is of paramount importance. The quality of a formulation and its delivery system determines the absorption and effectiveness. It's essential that you know your product and your supplier and above all observe your body's response to the product of your choice.

It's important, when companies suddenly appear on the market, that we do our homework. Many would-be entrepreneurs are just now hearing about the significance of the wild yam extract and are jumping on what they consider a get-rich-quick bandwagon. Some of these organizations are not using the essential, natural ingredients that are necessary to achieve the results we speak about in this book. A company that will evaluate the ingredients in your product (milligrams per ounce) is Scientific Associates, Inc., (314) 487-6776.

Incidentally, James Jamieson, a pharmaceutical manufacturer and researcher, says that some companies' products may function more naturally than others as a result of enzymatic fermentation of the yam rather than actual pharmaceutical alteration. Thousands of years' experience, he says, has proven the effectiveness of the former process. Jamieson calls these yams modulators: they bring hormone levels to normal whether they are high or low. They do not build up in the body, and you cannot overdose on them. On the other hand, one manufacturer of so-called nutraceuticals for dietary use told me, in fact, that the more sophisticated the extraction processes for phytochemicals have become, the less effective these isolated plant fractions have proven to be.

Along these lines, the Reader's Digest's *Magic and Medicine of Plants* poses some very interesting questions that are relevant to our consideration of the use of any herbal products:

> Could the very purity of laboratory-isolated substances be a drawback? Do some natural plant medicines have ingredients that prevent dangerous side effects in human use? . . . Is there sometimes a synergistic effect when the whole plant, as opposed to just the purified chemicals derived from it, is used? . . . Is the action of a whole plant sometimes more than the sum of its chemical parts?[6]

One might, therefore, want to work with manufacturers and pharmacies who are concerned with valid questions such as these. The substances that are often filtered out of the plant material do contain natural enzymes, alkaloids, peptides, phytosterols, etc. that interact with what is considered the "active" ingredient found in plants such as the wild yam. "They are part of the healing plant chemistry," says Dr. Richard Schulze. "To isolate the 'active ingredient' ignores all the OTHER ingredients that make an herb work."[7]

Plant compounds used in medicinal formulations are valuable, reliable, and easily assimilated sources of necessary building blocks, including proteins, carbohydrates, minerals, fatty acids, tannins, and many vitamins. These often function as precursors for our hormones and prostaglandins, and some even contain naturally occurring antibiotic elements. Barbara Griggs writes in *Green Pharmacy*: "Man and plant are close biological kin: the lifeblood of the plant, its green chlorophyll, has a chemical structure almost identical to the haemoglobin which is the central constituent of human blood; where chlorophyll has a molecule of magnesium in its structural pattern, haemoglobin carries a molecule of iron."[8] Marcia Jones, Director of the PMS and Menopause Center at Dixie Health, carries this analogy further, explaining that just as chlorophyll is nearly identical to hemoglobin, so diosgenin (from *Dioscorea*) is very similar to the molecular structure of progesterone and other hormones.

As a matter of course, we should seek out products whose manufacturers can assure us that the plants they use have been organically grown. Otherwise, residues of pesticides and other chemicals present in the end product could function as pseudo-estrogens and have the opposite from the desired effect, destroying hormonal balance.

Whatever the production method, it's worth mentioning that before purchasing any progesterone product, it pays to investigate whom you're buying from and what you are buying. The quality and potency of the ingredients can make a difference. And remember, each of us may have a unique way of responding to any particular progesterone formulation. Some women feel the difference within a couple of weeks, others several months. Again, what works for one person may not work as well for another—and it may be that all it takes is trying another brand, or using it long enough to give it a chance. And, as always is the case, if any unusual symptoms develop, consider consulting a health-care practitioner knowledgeable in the use of natural hormone replacement therapy. Certain modifications in use are probably necessary to suit your own body's needs.

Useful Reading

Guide to Women's Nutrition, by Carlton Fredericks, Ph.D., New York: The Putnam Publishing Group, 1989.

Health and Healing: Tomorrow's Medicine Today newsletters by Julian Whitaker, M.D.: Vol. 2, No. 7, July 1992; Vol. 3, No. 6, June 1993; Vol. 3, No. 3, March 1993; Vol. 4, No. 7, May 1994.

Hormone Replacement Therapy: Yes or No? by Betty Kamen, Ph.D., Bel Marin Keys, CA: Nutrition Encounter, Inc., 1991.

Menopaws by Martha Sacks, Berkeley, CA: 1994.

Natural Progesterone: The Multiple Roles of a Remarkable Hormone, by John R. Lee, M.D., Sebastopol, CA: BLL Publishing, 1993.

Naturally Well, newsletter by Marcus Laux, N.D.: Vol. 2, No. 12, December 1995.

Nutrition for Women (Fifth Edition), by Raymond F. Peat, Ph.D., Eugene, OR: Kenogen, 1993.

Once a Month, by Katharina Dalton, F.R.C.G.P., Pomona, CA: Hunter House, 1979.

PMS, Premenstrual Syndrome and You: Next Month Can Be Different, by Niels H. Lauersen, M.D., New York: Simon & Schuster, 1983.

Preventing and Reversing Osteoporosis, by Alan R. Gaby, M.D., Rocklin, CA: Prima Publishing, 1994.

The Silent Passage: Menopause, by Gail Sheehy, New York: Pocket Books, 1993.

What Your Doctor May Not Tell You about Menopause, by John R. Lee, M.D., with Virginia Hopkins, New York: Warner Books, 1996.

Without Estrogen: Natural Remedies for Menopause and Beyond, by Dee Ito, New York: Carol Southern Books, 1994.

Women's Bodies, Women's Wisdom, by Christiane Northrup, M.D., New York: Bantam Books, 1994.

Notes

1. Marcus Laux and Christine Conrad, *Natural Woman, Natural Menopause* (New York: HarperCollins, 1997), 70.

2. Jane E. Brody, *New York Times* National Section, November 28, 1994.

3. John R. Lee, *Natural Progesterone: The Multiple Roles of a Remarkable Hormone* (Sebastopol, CA: BLL Publishing, 1993), 76.

4. Ibid.

5. John R. Lee, *What Your Doctor May Not Tell You About Menopause* (New York: Warner Books, 1996), 270, 305.

6. Reader's Digest Editors, *Magic and Medicine of Plants* (New York: Reader's Digest Association, Inc., 1993), 71.

7. Sam Biser, *The Last Chance Health Report*, Vol. 6, Nos. 2–4, 12.

8. Barbara Griggs, *Green Pharmacy: The History and Evolution of Western Herbal Medicine* (New York: Viking Press, 1981), 332, 361.

NOTES

INTRODUCTION

1. Helen Keller, *Light in My Darkness*, revised and edited by Ray Silverman (Westchester, PA: Swedenborg Foundation, 1994).
2. John R. Lee, *Natural Progesterone: The Multiple Roles of a Remarkable Hormone* (Sebastopol, CA: BLL Publishing, 1993).
3. John R. Lee, "Is Natural Progesterone the Missing Link in Osteoporosis Prevention and Treatment?" *Medical Hypotheses*, 1991.
4. Julian Whitaker, *Health & Healing: Tomorrow's Medicine Today*, Vol. 3, No. 6, June 1993.
5. P. A. Lehmann, "Russell E. Marker," *Journal of Chemical Education*, Vol. 50, March 1973, 195–199.
6. Katharina Dalton, "Ante-natal Progesterone and Intelligence," *British Journal of Psychiatry*, Vol. 114, 1968.
7. A. H. Follingstad, "Estriol, the Forgotten Estrogen?" *Journal of the American Medical Association*, Vol. 239, No. 1, January 2, 1978.
8. Joel T. Hargrove et al., "Menopausal Hormone Replacement Therapy with Continuous Daily Oral Micronized Estradiol and Progesterone," *Obstetrics & Gynecology*, Vol. 73, No. 4, April 1989, 606–612.
9. J. C. Prior, Y. Vigna, and N. Alojada, "Progesterone and the Prevention of Osteoporosis," *Canadian Journal of OB/Gyn & Women's Health Care*, Vol. 3, No. 4, 1991, 181.
10. John R. Lee, "Significance of Molecular Configuration Specificity—The Case of Progesterone and Osteoporosis," *Townsend Letter for Doctors*, June 1993, 558.
11. John R. Lee, *Natural Progesterone*.
12. John R. Lee, "Significance of Molecular Configuration Specificity," 558.
13. John R. Lee, "Natural Progesterone" (transcript), *Cancer Forum*, Vol. 13, No. 5/6, Winter 1994–1995.
14. Keller, *Light in My Darkness*.

CHAPTER 1: SICK AND TIRED OF BEING TIRED AND SICK

1. Macmillan Dictionary of Quotations (New York: Macmilllan, 1989), 358.
2. Richard A. Passwater, *Cancer and Its Nutritional Therapies* (New Canaan, CT: Keats Publishing, 1983).
3. Raquel Martin, *Today's Health Alternative* (Bozeman, MT: America West Publishers), 1992.
4. "Provera, brand of medroxyprogesterone acetate tablets," USP Upjohn Company, Kalamazoo, MI, 1992.
5. Gail Sheehy, *The Silent Passage: Menopause* (New York: Pocket Books, 1993).
6. John R. Lee, "Is Natural Progesterone the Missing Link in Osteoporosis Prevention and Treatment?" *Medical Hypotheses*, 1991.
7. John R. Lee, *Natural Progesterone: The Multiple Roles of a Remarkable Hormone* (Sebastapol, CA: BLL Publishing, 1993).
8. John R. Lee, *What Your Doctor May Not Tell You About Menopause* (New York: Warner Books, 1996), 257.
9. John R. Lee, "Natural Progesterone," *Cancer Forum*, Vol. 13, No. 5/6, 1994–1995, 9.
10. John R. Lee, *What Your Doctor May Not Tell You About Menopause*, 89, 197, 254.
11. Keller, *Light in My Darkness*, revised and edited by Ray Silverman (Westchester, PA: Swedenborg Foundation, 1994).
12. Ibid.
13. Sandra Coney, *The Menopause Industry: How the Medical Establishment Exploits Women* (Alameda, CA: Hunter House, 1994), 270.
14. John R. Lee, *Natural Progesterone*.
15. John R. Lee, "Is Natural Progesterone the Missing Link in Osteoporosis Prevention and Treatment?" *Medical Hypotheses*, 1991.
16. John R. Lee, *Natural Progesterone*, iii, 33, 84–86.
17. Ibid., 46, 47.
18. Ibid., 26, 44, 45.
19. Ibid., iii, 4, 71–75, and John R. Lee, *What Your Doctor May Not Tell You About Menopause*, 207–214.

CHAPTER 2: PROGESTERONE DEFICIENCY, YES; ESTROGEN, MAYBE . . .

1. John R. Lee, *Natural Progesterone: The Multiple Roles of a Remarkable Hormone* (Sebastopol, CA: BLL Publishing, 1993), 55.
2. Ibid., 5–6.
3. John R. Lee, *What Your Doctor May Not Tell You About Menopause* (New York: Warner Books, 1996).
4. Lita Lee, "Estrogen, Progesterone and Female Problems," *Earthletter*, Vol. 1, No. 2, June 1991.
5. Raymond F. Peat, "Progesterone: Essential to Your Well-Being," *Let's Live*, April 1982.
6. John R. Lee, "Is Natural Progesterone the Missing Link in Osteoporosis Preven-

tion and Treatment?" *Medical Hypotheses*, 1991.

7. Peter S. Rhodes, *AIM* (Bryn Athyn, PA: New Will, 1991).

8. Gail Sheehy, *The Silent Passage: Menopause*, (New York: Pocket Books, 1993), 29.

9. Lila Natchtigall and John Heilman, *Estrogen: The Facts Can Change Your Life* (Los Angeles: The Body Press, 1986).

10. Sandra Coney, *The Menopause Industry: How the Medical Establishment Exploits Women* (Alameda, CA: Hunter House, 1994), 77.

11. Ibid., 195.

12. Ibid., 199.

13. Ibid., 198.

14. Ibid., 21–22, 213.

15. John R. Lee, *Natural Progesterone*, 8, 54, 74.

16. Lita Lee, "Estrogen, Progesterone, and Female Problems."

17. Alvin Follingstad, "Estriol, the Forgotten Estrogen?" *Journal of the American Medical Association*, Vol. 239, No. 1, January 2, 1978.

18. Julian Whitaker, *Health & Healing*, Vol. 3, No. 3, March 1993.

19. Graham A. Colditz, "Type of Post-Menopausal Hormone Use and Risk of Breast Cancer 12-Year Follow Up from the Nurses' Health Study," *Cancer Causes and Control*, Vol. 3, Sept. 1992, 433–439.

20. M. A. J. McKenna, "Breast Cancer Risk Tied to Years of Menopausal Hormone Care," *Atlanta Journal/Constitution*, June 15, 1995.

21. Lorrain Dusky, "Progesterone: Safe Antidote for PMS (Women's Health Report)," *McCall's*, October 1990.

22. Ibid.

23. Ibid.

24. I. Wiklund et al., "Long-Term Effect of Transdermal Hormonal Therapy on Aspects of Quality of Life in Postmenopausal Women," *Maturitas*, Vol. 3, March 14, 1992, 225–236.

25. Dusky, "Progesterone."

26. John R. Lee, *Natural Progesterone*, and *What Your Doctor May Not Tell You About Menopause*.

27. Niels H. Lauersen, *PMS, Premenstrual Syndrome and You: Next Month Can Be Different* (New York: Simon & Schuster, 1983), 170.

28. John R. Lee, *Natural Progesterone*.

29. Ibid., 2.

30. Ibid., 35, 41, 54.

31. Lita Lee, "Estrogen, Progesterone, and Female Problems."

32. Katharina Dalton, "Premenstrual Syndrome and Postnatal Depression," *Health and Hygiene*, Vol. 11, 1990, 199–201.

33. Betty Kamen, *Hormone Replacement Therapy: Yes or No?* (Novato, CA: Nutrition Encounter, Inc., 1993), 4, 104, 210.

34. John R. Lee, *What Your Doctor May Not Tell You About Menopause*, 254.

35. Ibid., 197.

36. Ibid., 89.

37. Ibid., 24–33, 43–49.

38. John R. Lee, *Natural Progesterone* 34, 35, 58, 65–70.
39. Alan R. Gaby, *Preventing and Reversing Osteoporosis* (Rocklin, CA: Prima Publishing, 1994), 9, 10, 21–22.
40. Raquel Martin, *Today's Health Alternative.*
41. John R. Lee, *Natural Progesterone,* 4–8.
42. Lita Lee, "Estrogen, Progesterone, and Female Problems."
43. Peat, "Progesterone."
44. Kamen, *Hormone Replacement Therapy,* 210.
45. Ibid., 210.
46. John R. Lee, "Significance of Molecular Configuration Specificity."
47. John R. Lee, *Natural Progesterone.*
48. John R. Lee, *What Your Doctor May Not Tell You About Menopause,* 270.
49. Peat, *Nutrition for Women,* 5th ed. (Eugene, OR: Kenogen, 1993), 84–85.
50. "Natural Progesterone/Body Creams Vary in Potency," *Women's Health Advocate,* Vol. 3, No. 7, 1996, 7.
51. John R. Lee, *Natural Progesterone.*
52. John R. Lee, *Natural Progesterone,* 79.
53. Christiane Northrup, *Women's Bodies, Women's Wisdom* (New York: Bantam Books, 1994), 76.
54. Peat, "Effectiveness of Progesterone Assimilation for the Relief of Premenstrual Symptoms" [educational brochure]. Eugene, OR.
55. Joel T. Hargrove et al., "Absorption of Oral Progesterone Is Influenced by Vehicle and Particle Size," *American Journal of Obstetrics and Gynecology,* Vol. 161, No. 4, October 1989, 948–951.
56. Ibid.
57. Rabbi Eric Braverman, "Most Asked Questions About Natural Oral Progesterone," *Total Health,* Vol. 17, No. 2, April 1995.
58. Peat, *Progesterone in Orthomolecular Medicine* (Eugene, OR: Kenogen, 1993).
59. Joel T. Hargrove et al., "Absorption of Oral Progesterone."
60. Joel T. Hargrove et al., "Menopausal Hormone Replacement Therapy," 606–612.
61. Ibid.
62. Ibid.
63. Ibid.
64. Ibid.
65. Hargrove et al., "Absorption of Oral Progesterone."
66. Peat, *Progesterone in Orthomolecular Medicine.*
67. J. Jensen et al., "Long-term Effects of Percutaneous Estrogens and Oral Progesterone on Serum Lipoproteins in Postmenopausal Women," *American Journal of Obstetrics and Gynecology,* Vol. 156, 1987, 66–71.
68. Peat, *Progesterone in Orthomolecular Medicine,* 46, 63.
69. Hargrove et al., "Absorption of Oral Progesterone."
70. Hargrove et al., "Menopausal Hormone Replacement Therapy."
71. "Specific Estrogen Treatment for Atrophy Related Urogenital Complaints" (Ovestin estriol scientific information), The Netherlands, Organon International.
72. A. Philips, D. W. Hahn, and J. L. McGuire, "Comparative Effect of Estriol and

Equine Conjugated Estrogens on the Uterus and the Vagina," *Marturitas*, Vol. 5, 1984, 147–152.

73.   I. Milsom et al. "Vaginal Immunoglobulin A (IgA) levels in Post-Menopausal Women: Influence of Oestriol Therapy," *Maturitas*, Vol. 13, 1991, 129–135.

74.   A. Brandberg, D. Mellstrom, and G. Samsioe, "Low Dose Estriol Treatment in Elderly Women with Urogenital Infections," *Acta Obstet. Scand.* (supplement), Vol. 140, 1987, 33–38.

75.   G. M. Heimer, "Estriol in the Postmenopause," *Acta Obstet. Gynecol. Scand.* (supplement), Vol. 139, 1987, 5–23.

76.   A. L. Kirkengen et al., "Oversterin (Estriol) in the Prophylactic Treatment of Recurrent Urinary Infections in Postmenopausal Women," *European Journal of Pharmacology*, Vol. 183, 1990, 1040.

77.   C. Zara et al., "Estriol in the Treatment of Postmenopausal Osteoporosis," in Proceedings of the Second European Winter Conference in Gynecology and Obstetrics; Italy, March 1989, ed. A. R. Genazzani, F. Petragli, A. Volpe, *Cornforth* (Parthenon Publishing, 1990), 925–929.

78.   L. A. Mattsson and G. Cullberg, "A Clinical Evaluation of Treatment with Estriol Vaginal Cream Versus Suppository in Postmenopausal Women," *Acta Obstet. Gynecol. Scand.*, Vol. 62, 1983, 397–410.

79.   J. Lambillon, "L'Aacifemine Companee a l'Equigyne Dans la Therapeutique des Troubles de la Menopause" (Aacifemine Compared with Equigyne in the Treatment of Climacteric Complaints), *Bull. Soc. R. Belge. Gynecol. Obstet.*, Vol. 39, 1969, 135–146.

80.   P. M. Kicovic et al., "The Treatment of Postmenopausal Vaginal Atrophy with Ovestin Vaginal Cream or Suppositories: Clinical, Endocrinological and Safety Aspects," *Maturitas*, Vol. 2, 1980, 275–282.

81.   Ibid.

82.   V. A. Tzingounis, M. Feridun Aksu, and R. B. Greenblatt, "The Significance of Oestriol in the Management of the Postmenopause," *Acta Endocrinologica* (supplement), Vol. 233, 1980, 45–50.

83.   J. Proost et al., "Oestriolbehandeling Van De Menopauze En Postmenopauze" (Estriol Treatment in Menopause and Postmenopause), *Ned. Tijdschr. Geneeskd.*, Vol. 2, 1969, 94–99.

84.   R. C. Merrill, "Estriol: A Review," *Physiology Reviews*, Vol. 38, No. 3, 1958, 463–480, and Peat, *Ray Peat's Newsletter*, Sept. 1996.

85.   Julian Whitaker, *Health and Healing*, Vol. 7, No. 6, June 1997, 7.

86.   Tzingounis et al., "The Significance of Oestrol," and Proost et al., "Oestriol behandeling."

87.   G. Warnecke, "Die Behandlung Des Menopausesyndroms Mit Oestriol" (The Treatment of the Climacteric Syndrome with Estriol), *Z. Ther.*, Vol. 7, 1972, 423–436.

88.   Ibid.

89.   Tzingounis et al., "The Significance of Oestriol."

90.   Proost et al., "Oestriolbehandling."

91.   Ibid.

92. "Specific Estrogen Treatment for Atrophy Related Urogenital Complaints."

93. E. W. Bergink et al., "Effect of Oestriol, Oestradiol Valerate and Ethinyloestradiol on Serum Proteins in Oestrogen-Deficient Women," *Maturitas*, Vol. 3, 1981, 241–247.

94. Gaby, *Preventing and Reversing Osteoporosis*, 132.

95. Ibid.

96. Lita Lee, "Estrogen."

97. Marcus Laux, *Naturally Well*, Vol. 2, No. 12, December 1995.

98. Ibid.

99. Sharon Gleason, "Menopause . . . It's Not a Disease," *The Rice Paper*, Summer 1994.

100. Coney, *The Menopause Industry*.

101. U. B. Ottosson, "Oral Progesterone and Estrogen/Progestogen Therapy: Effects of Natural and Synthetic Hormones on Subfractions of HDL Cholesterol and Liver Proteins," *Acta Obstet. Gynecol. Scand.* (supplement), Vol. 127, 1984, 1–37.

102. John R. Lee, *Natural Progesterone*, 8.

103. Ibid., 32–37, 71–75.

104. Ibid., 33.

105. Peat, *Nutrition for Women* (5th ed.), (Eugene, OR: 1993).

106. Ibid.

107. John R. Lee, *Natural Progesterone*, 84.

108. Ibid., 70, 82, 85, 86.

109. The Burton Goldberg Group, *Alternative Medicine* (Puyallup, WA: Future Medicine Publishing, 1993), 675.

110. Nina Sessler, "Questions & Answers on Women's Health Issues: What Is Fibrocystic Breast Disease?" *Natural Solutions*, Vol. 1, Issue 1, Fall 1993.

111. Susan M. Love, *Dr. Susan Love's Breast Book* (New York: Addison-Wesley, 1990), 81–86.

112. John R. Lee, *Natural Progesterone*, 84.

113. Ibid., 87.

114. Ibid., 87.

115. Gaby, *Preventing and Reversing Osteoporosis*, 154.

116. Majid Ali, "Chemical Conflict: The Age of Estrogen Overdrive," *Lifespanner*, December 1994.

117. Linda G. Rector-Page, *Healthy Healing: An Alternative Healing* (Soquel, CA: Healthy Healing Publications, 1992).

118. Anne Dickson and Nikki Henriques, *Women on Menopause* (Rochester, VT: Healing Arts Press, 1988), 55.

119. Ellen Brown and Lynne Walker, *Breezing Through the Change*, (Berkeley, CA: Frog, Ltd. 1994), 77.

120. Sheehy, *The Silent Passage*.

121. Ibid.

122. Ibid.

123. Ibid.

124. Dickson and Henriques, *Women on Menopause*, 6.

125. Ibid.
126. John R. Lee, *Natural Progesterone,* 30.
127. Peat, "Progesterone."
128. John R. Lee, *What Your Doctor May Not Tell You About Menopause,* 89, 197, 198, 254.
129. John R. Lee, *Natural Progesterone,* 33.
130. Kamen, *Hormone Replacement Therapy,* 210.
131. Ibid., 210.
132. Peat, "Estrogen: Simply Dangerous," *Ray Peat's Newsletter,* July 1995.

CHAPTER 3: THE SEASONS OF A WOMAN'S LIFE

1. José Ortega y Gasset, *The Revolt of the Masses* (1930).
2. Sterling Morgan, "Part II: PMS, Menopause, & Other Areas," *To Your Health,* September/October 1993, 19.
3. Katharina Dalton, *Once a Month* (Alameda, CA: Hunter House, 1994).
4. Stuart M. Berger, *What Your Doctor Didn't Learn in Medical School* (New York: William Morrow, 1988), 211–213.
5. Harriet Greveal, "Answers for PMS," *Total Health,* July 1985, 26.
6. Chakmakjian Zaren, "A Critical Assessment of Therapy for the Pre-Menstrual Tension Syndrome," *Journal of Reproductive Medicine,* Vol. 28, No. 8, Aug. 1983, 532.
7. Dalton, "Premenstrual Syndrome and Postnatal Depression," *Health and Hygiene,* Vol. 11, 1990, 199–201.
8. Dalton, "Guide to Progesterone for Postnatal Depression" (pamphlet), 1990.
9. Ibid.
10. Dalton, *A Guide to Premenstrual Syndrome and Its Treatment* (pamphlet), 1990, 12.
11. Dalton, "Premenstrual Syndrome."
12. John R. Lee, *Natural Progesterone: The Multiple Roles of a Remarkable Hormone* (Sebastopol, CA: BLL Publishing, 1993), 33.
13. John R. Lee, "Natural Progesterone" (transcript), *Cancer Forum,* Vol. 13, No. 5/6, Winter 1994-1995, 61.
14. Betty Kamen, *Hormone Replacement Therapy: Yes or No?* (Novato, CA: Nutrition Encounter, Inc., 1993), 212.
15. Sterling Morgan, "Pregnancy & Natural Progesterone = Superior Baby," *To Your Health: The Magazine of Healing and Hope,* Vol. V, No. 6, May/June/July 1993.
16. Raymond F. Peat, *Nutrition for Women,* 5th ed. (Eugene, OR: Kenogen, 1993).
17. Dalton, *Once a Month,* 53.
18. Kamen, *Hormone Replacement Therapy,*
19. Niels H. Lauersen, *PMS: Premenstrual Syndrome and You—Next Month Can Be Different* (New York: Simon & Schuster, 1983).
20. Gail Sheehy, *The Silent Passage: Menopause* (New York: Pocket Books, 1993).
21. Peat, *Nutrition for Women,* 18.
22. Morgan, "Part II: Menopause."

23. Morgan, "Pregnancy and Natural Progesterone."
24. Ibid.
25. Peat, "Progesterone: Essential to your Well-Being," *Let's Live*, April 1982.
26. Lee, *Natural Progesterone*, 29.
27. Ibid., 30.
28. J. C. Prior and Y. M. Vigna, "Spinal Bones Loss and Ovulatory Disturbances," *New England Journal of Medicine*, Vol. 223, 1990, 1221–1227.
29. J. C. Prior, Y. M. Vigna, and N. Alojada, "Progesterone and the Prevention of Osteoporosis," *Canadian Journal of OB/GYN & Women's Health Care*, Vol. 3, 1991, 178–184.
30. Peat, "Progesterone."
31. Norman C. Shealey, *DHEA: The Health and Youth Hormone* (New Canaan, CT: Keats Publishing, 1996).
32. Alan R. Gaby, *Preventing and Reversing Osteoporosis* (Rocklin, CA: Prima Publishing, 1994), 165–168.
33. "Mother Knows Best! The Fascinating Healing Benefits of DHEA—The Body's "Mother Hormone!," *Bio/Tech News*, Portland, OR, 1995, 5.
34. Shealy, *DHEA*.
35. A. J. Morales et al., "Effects of Replacement Dose of Dehydroepiandrosterone in Men and Women of Advancing Age," *Journal of Clinical Endocrinology and Metabolism*, Vo. 78, No. 6, 1994, 1360–1367.
36. Joe Glickman, Jr., "Hormone Holds Key to Aging Process," *Health Science Report*, Vol. 2, No. 1, 1996.
37. Elizabeth Barrett-Conner et al., "A Prospective Study of Dehydroepiandrosterone Sulfate, Mortality, and Cardiovascular Disease," *New England Journal of Medicine*, Vol. 315, No. 24, 1986, 1519–24.
38. D. M. Herrrington, G. B. Gordon, et al., "Plasma Dehydroepiandrosterone and Dehydroepiandrosterone Sulfate in Patients Undergoing Diagnostic Coronary Angiography," *Journal of the American College of Cardiology*, Vol. 16, No. 6, 1990, 862–70.
39. J. E. Nestler et al., "Dehydroepiandrosterone Reduces Serum Low Density Lipoprotein Levels and Body Fat But Does Not Alter Insulin Sensitivity in Normal Men," *Journal of Clinical Endocrinology and Metabolism*, Vol. 66, No. 1, 1988, 57–61.
40. G. B. Gordon et al., "Reduction of Atherosclerosis by Administration of Dehydroepiandrosterone. A Study in the Hypercholesteroemic New Zealand White Rabbit With Aortic Intimal Injury," *Journal of Clinical Investigations*, Vol. 82, No. 2, 1988, 712–720.
41. Glickman, "Hormone Holds Key to Aging Process."
42. G. B. Gordon, "Reduction of Atherosclerosis."
43. A. G. Schwartz and L. L. Pashko, "Cancer Chemoprevention with the Adrenocortical Steroid Dehydroepiandrosterone and Structural Analogs," *Journal of Cellular Biochemistry* (Supplement), Vol. 17G, 1993, 73–79.
44. G. B. Gordon et al., "Serum Levels of Dehydroepiandrosterone Sulfate and the Risk of Developing Gastric Cancer," *Cancer Epidemiology*, Vol. 2, No. 1, 1993, 33–35.

45. G. B. Gordon, L. M. Shantz, and P. Talalay, "Modulation of Growth, Differentiation and Carcinogenesis by Dehydroepiandrosterone," *Advances in Enzyme Regulation*, Vol. 26, 1987, 355–382.

46. C. W. Boone, V. E. Steele, and G. J. Kelloff, "Screening for Chemopreventive (Anticarcinogenic) Compounds in Rodents," *Mutation Research*, Vol. 267, No. 2, 1992, 251–255.

47. L. Bologa, J. Sharma, and E. Roberts, "Dehydroepiandrosterone and its Sulfated Derivative Reduce Neuronal Death and Enhance Astrocytic Differentiation in Brain Cell Cultures," *Journal of Neuroscience Research*, Vol. 17, No. 3, 1987, 225–234.

48. Glickman, "Hormone Holds Key."

49. Glickman, "Hormone Holds Key to Aging Process," 3.

50. Peat, "Progesterone."

51. Alfred Gilman, Louis Goodman, *Goodman & Gilman's Pharmacological Basis of Therapeutics* (6th ed.) (New York: Macmillan, 1980), 1420–1438.

52. Sheehy, *The Silent Passage.*

53. Ibid.

54. Ibid.

55. Ibid.

56. Bert Stern et al., *The Pill Book* (New York: Bantam Books, 1986), 542.

57. D. T. Baird and A. F. Glasier, "Drug Therapy: Hormonal Contraception," *New England Journal of Medicine*, Vol. 328, No. 1543, 1993, 1543.

58. Kamen, *Hormone Repacement Therapy*, 35.

59. Anne Dickson and Nikki Henriques, *Women on Menopause* (Rochester, VT: Healing Arts Press, 1988), 67.

60. Lita Lee, "Estrogen, Progesterone and Female Problems," *Earthletter*, Vol. 1, No. 2, June 1991.

61. Carlton Fredericks, *Guide to Women's Nutrition* (New York: Putnam Publishing Group, 1989).

62. John R. Lee, *Natural Progesterone*, 44.

63. Ibid.

64. Julian Whitaker, *Health & Healing Tomorrow's Medicine Today*, Vol. 4, No. 5, May 1994.

65. Ibid.

66. The Burton Goldberg Group, *Alternative Medicine* (Puyallup, WA: Future Medicine Publishing, 1993), 1014–1015.

67. Ibid.

68. Joe Graedon, *The People's Pharmacy* (New York: Avon Books, 1977),188–193.

69. "Dangers of Estrogen," Form P8263-01, Mead Johnson Laboratories, Princeton, NJ, December 1992.

70. John R. Lee, *What Your Doctor May Not Tell You About Menopause* (New York: Warner Books, 1996), 307.

71. Raquel Martin, *Today's Health Alternative* (Bozeman, MT: America West Publishers, 1992).

72. Sheehy, *The Silent Passage*, 124.

73. Donald J. Brown, *Herbal Prescriptions for Better Health* (Rocklin, CA: Prima, 1996), 181, 182, and D. Propping et al., "Diagnosis and Therapy of Corpus Luteum Insufficiency in General Practice," *Therapiewoche*, Vol. 38, 1988, 2992–3001.

74. Peat, *Nutrition for Women*, 16, 18.

75. Ibid., 16.

76. Mark Perloe and Linda Gail, *Miracle Babies and Other Happy Endings for Couples with Fertility Problems* (Atlanta, 1986, http://www.ivf.com//ch12mb.html).

77. Peat, *Progesterone in Orthomolecular Medicine* (Eugene, OR: Kenogen 1995).

78. Peat, *Nutrition for Women*, ii–iv, 17, 116.

79. Peat, "Stress and Water," *Ray Peat's Newsletter*, February 1995.

80. Mohammed Kalimi and William Regelson, eds., *The Biologic Role of Dehydroepiandrosterone (DHEA)* (Hawthorne, NY: Walter de Gruyter, 1990).

81. Peat, *Nutrition for Women*, 43.

82. John R. Lee, *What Your Doctor May Not Tell You About Menopause*, 252–254.

83. Richard P. Huemer, "Fibromyalgia: The Pain That Never Stops," *Let's Live*, November 1996, 34–35.

84. Betty Kamen, "Fibromyalgia: An Age-Old Malady Begging for Respect," *Let's Live*, November 1994, 31.

85. Ibid., 32, and *Journal of Internal Medicine*, Vol. 235, March 1994, 199.

86. Sandra Coney, *The Menopause Industry* (Alameda, CA: Hunter House, 1994), 61.

87. Colditz, K. M. Egan, and M. J. Stampfer, "Hormone Replacement Therapy and Risk of Breast Cancer: Results from Epidemiologic Studies," *American Journal of Obstetrics and Gynecology*, Vol. 168, 1993, 1476–1480.

88. Quoted in Ann Louise Gittleman, "Menopause and Nutrition," *The Energy Times*, September/October 1994.

89. Anthony J. Chichoke, *The Energy Times*, May/June 1994.

90. Gilman and Goodman, *Goodman and Gilman's Pharmacological Basis.*

91. Ibid.

92. John R. Lee, *Natural Progesterone*, 40–41.

93. Shealy, *DHEA.*

94. Christine Ammer, *The New A-Z of Womens' Health: A Concise Encyclopedia* (New York: Facts on File, Inc., 1989), 357.

95. Coney, *The Menopause Industry.*

96. Dalton, "Guide to Progesterone."

97. Peat, "Progesterone."

98. Linda G. Rector-Page, *Healthy Healing: An Alternative Healing Reference,* (Soquel, CA: Healthy Healing Publications, 1992), 166.

99. Raymond Peat, "Diabetes, Scleroderma, Oils and Hormones," *Ray Peat's Newsletter*, Issue 131, July 1995.

100. Lee, *Natural Progesterone*, 43.

101. Carol Landau, Michele G. Cyr, and Anne W. Moutlon, *The Complete Book of Menopause* (New York: Putnam Books, 1994), 153.

102. Gaby, *Preventing and Reversing Osteoporosis*, 149–150.

103. John R. Lee, *Natural Progesterone*, 46.

104. Lita Lee, "Estrogen."
105. Dalton, *A Guide to Premenstrual Syndrome.*
106. Dalton, "Guide to Progesterone."
107. Lindacarol Graham, "Do you have a Hormone Shortage?" *Redbook,* February 1989, 16.
108. Jerome H. Check and Harriet G. Adelson, "The Efficacy of Progesterone in Achieving Successful Pregnancy: II. In Women with Pure Luteal Phase Defects," *International Fertility,* Vol. 32, No. 21, 139–141.
109. Ibid.
110. Ibid.
111. Tzinounis, S. Michalas, and D. Kaskarelis, "Effect of Oestriol on Cervical Mucus," *Clinical Trials Journal,* Vol. 19, 1982, 3844; "Specific Estrogen Treatment for Atrophy Related Urogenital Complaints" (Ovestin estriol scientific information). Product Surveillance department, Organon International, The Netherlands.
112. Christiane Northrup, *Health Wisdom for Women,* Vol. 2, No. 11, November 1995, 6; October 1995.
113. Dalton, "Ante-natal Progesterone and Intelligence," *British Journal of Psychiatry,* Vol. 114, 1968, 1377–1382.
114. Ibid.
115. Dalton, *Once a Month,* 163–164.
116. Ibid., 169, 232; "Premenstrual Syndrome"; and "Prenatal Progesterone and Educational Attainments," *British Journal of Psychiatry,* Vol. 129, 1976, 438–442.
117. Greveal, "Answers for PMS."
118. Dalton, *Once a Month,* 161–169.
119. John R. Lee, *Natural Progesterone,* 81, 87.
120. John R. Lee, *What Your Doctor May Not Tell You About Menopause,* 80.
121. Gaby, *Preventing and Reversing Osteoporosis,* 148.
122. Dalton, *Once a Month,* 163.
123. Gaby, *Preventing and Reversing Osteoporosis,* 261.
124. McKenna, "Darwin with a Twist," *The Atlanta Journal/Constitution,* April 25, 1995.
125. Peat, "Estrogen Simply Dangerous," *Ray Peat's Newsletter,* Eugene, OR, July 1995, 3.
126. Lita Lee, "Estrogen."
127. Peat, *Nutrition for Women.*
128. Peat, "Progesterone."
129. Peat, "The Progesterone Deception," *Townsend Letter for Doctors,* November 1987.
130. Peat, *Progesterone in Orthomolecular Medicine.*
131. John R. Lee, "Estrogen."
132. Morgan, "Pregnancy and Natural Progesterone."
133. John R. Lee, *Natural Progesterone,* 25.
134. Ibid., 17, 41.
135. Dalton, "Guide to Progesterone."

136. Ibid.
137. Ibid.
138. Stern et al., *The Pill Book*, 351.
139. Dalton, "Ante-natal Progesterone."

CHAPTER 4: CAN WE CIRCUMVENT OSTEOPOROSIS?

1. Joseph E. Maynard, *Healing Hands* (Woodstock, GA: Jonorm Publishing, 1991).
2. John R. Lee, *Natural Progesterone: The Multiple Roles of a Remarkable Hormone* (Sebastopol, CA: BLL Publishing, 1993), 56.
3. Ibid.
4. Sandra Cabot, *Smart Medicine for Menopause* (Garden City Park, NY: Avery Publishing, 1995), 30.
5. John R. Lee, *Natural Progesterone*, 56.
6. J. C. Prior, "Progesterone as a Bone-Trophic Hormone," *Endocrine Reviews*, Vol. 11, No. 2, 1990, 386–398.
7. John R. Lee, *Natural Progesterone*, 32.
8. Ibid., 30, 59–69, and Lita Lee, "Estrogen, Progesterone and Female Problems," *Earthletter*, Vol. 1, No. 2, June 1991.
9. Leslie Lawrence, "Education Can Ward off Osteoporosis," *The Atlanta Journal/ The Atlanta Constitution*, September 14, 1993.
10. *The Holistic Dental Digest*, No. 76, June 1992.
11. M. J. Halberstam, "If Estrogens Retard Osteoporosis, Are They Worth the Cancer Risk?" *Modern Medicine*, Vol. 45, No. 9, 1977, 15.
12. Gail Sheehy, *The Silent Passage: Menopause* (New York: Pocket Books, 1993), 180–181.
13. Ibid., 125–126.
14. J. C. Prior, Y. Vigna, and N. Alojada, "Spinal Bone Loss and Ovulatory Disturbances," *International Journal of Gynecology and Obstetrics*, Vol. 34, 1990, 253–256.
15. John R. Lee, "Is Natural Progesterone the Missing Link in Osteoporosis Prevention and Treatment?" *Medical Hypotheses*, 1991.
16. Majid Ali, "Chemical Conflict: The Age of Estrogen Overdrive," *Lifespanner*, December 1994.
17. Sheehy, *The Silent Passage*, 181–182.
18. Alan R. Gaby, *Preventing and Reversing Osteoporosis* (Rocklin, CA: Prima Publishing, 1994), 234.
19. Joel Griffiths, "Progesterone Reported to Increase Bone Density 10% in Six Months," *Medical Tribune*, November 29, 1990.
20. D. T. Felson et al., "The Effect of Postmenopausal Estrogen Therapy on Bone Density in Elderly Women," *New England Journal of Medicine*, Vol. 88, No. 8, Oct. 14, 1993.
21. John R. Lee, *Natural Progesterone*, 54–55.
22. D. R. Rudy, "Hormone Replacement Therapy," *Postgraduate Medicine*, December 1990, 157–164.

23. Gaby, *Preventing and Reversing Osteoporosis,* ix–x.
24. John R. Lee, *Natural Progesterone,* 54.
25. Ibid.
26. John R. Lee, "Osteoporosis Reversal: The Role of Progesterone," *International Clinical Nutrition Review,* Vol. 10, No. 3, 1990.
27. John R. Lee, "Fighting Osteoporosis with Natural Progesterone," *Natural Solutions,* Vol. 1, Issue 1 (Portland, OR: Professional & Technical Services Inc., Fall 1993).
28. John R. Lee, "Osteoporosis Reversal: The Role of Progesterone," 384–391.
29. Ibid.
30. Ibid.
31. Ibid.
32. John R. Lee, *Natural Progesterone,* 42, 54–55, and Prior, "Progesterone as a Bone Trophic Hormone," 386–398.
33. Sheehy, *The Silent Passage,* 183.
34. Alan Cook, "Osteoporosis: Review and Commentary," *Journal of the Neuro-musculoskeletal System,* Vol. 2, No. 1, 1994.
35. John R. Lee, *Natural Progesterone,* 85.
36. Gaby, *Preventing and Reversing Osteoporosis,* 143–152.
37. John R. Lee, "Is Natural Progesterone the Missing Link in Osteoporosis Prevention and Treatment?"
38. Ibid.
39. Ibid., and "Osteoporosis Reversal: The Role of Progesterone."
40. John R. Lee, "Is Natural Progesterone the Missing Link in Osteoporosis Prevention and Treatment?"
41. John R. Lee, "Osteoporosis Reversal: The Role of Progesterone."
42. John R. Lee, *Natural Progesterone,* 41, and "Is Natural Progesterone the Missing Link in Osteoporosis Prevention and Treatment?"
43. John R. Lee, *Natural Progesterone,* 59.
44. John R. Lee, "Is Natural Progesterone the Missing Link in Osteoporosis Prevention and Treatment?", and Sheehy, *The Silent Passage,* 98, 245.
45. Gaby, *Preventing and Reversing Osteoporosis,* 1–7.
46. Ibid.
47. Dee Ito, *Without Estrogen: Natural Remedies for Menopause and Beyond* (New York: Carol Southern Books, 1994), 6.
48. Joel Fuhrman, "Osteoporosis: How to Get It and How to Avoid It," *Health Science,* January/February 1992.
49. Ibid.
50. Ibid.
51. Prior, "Progesterone as a Bone-Trophic Hormone."
52. Fuhrman, "Osteoporosis: How to Get It and How to Avoid It."
53. M. T. Morter, Jr., "Osteoporosis!!" *The Chiropractic Professional,* May/June 1987, and *Your Health Your Choice: Guide to Nutrition and Disease Prevention* (Rogers, AK: B.E.S.T. Research, 1990), 3, 83, 167.
54. Anne Dickson and Nikki Henriques, *Women on Menopause* (Rochester, VT, Healing Arts Press, 1988), 78.

55. Rosemary Gladstar, "A New Cycle of Life: Celebrating Menopause Herbally," *The Herb Quarterly,* Winter 1993, 42.

56. Betty Kamen, *Hormone Replacement Therapy: Yes or No?* (Bel Marin Keys, CA: Nutrition Encounter, 1991).

57. Cathy Perlmutter and Toby Hanlon, "Triumph over Menopause," *Prevention,* August 1994.

58. John R. Lee, *Natural Progesterone,* 60.

59. Anthony J. Cochoke, "What Your Customer Should Know," *Health Food Business Magazine,* September 1996.

60. John R. Lee, *What Your Doctor May Not Tell You About Menopause* (New York: Warner Books, 1996), 296.

61. Ibid.

62. John A. McDougall and Mary A. McDougall, *The McDougall Plan* (Piscataway, NJ: New Century Publishers, Inc., 1983).

63. Gaby, *Preventing and Reversing Osteoporosis,* 6.

64. Ibid., 44.

65. John R. Lee, *Natural Progesterone,* 60.

66. Jason Elias and Katherine Ketcham, *In the House of the Moon* (New York: Warner Books, 1995), 292–293.

67. Gaby, *Preventing and Reversing Osteoporosis,* 108.

68. G. E. Abraham and H. Grewal, "A Total Dietary Program Emphasizing Magnesium Instead of Calcium: Effect on the Mineral Density of Calcaneous Bone in Postmenopausal Women on Hormonal Therapy," *Journal of Reproductive Medicine,* Vol. 35, 1990, 503–507.

69. Morton Walker, *The Chelation Way* (New York: Avery Publishing Group, 1990), 26.

70. M. L. Brandi, "Flavonoids: Biochemical Effects on Therapeutic Applications," *Bone and Mineral* 19 (Supplement), S3–S14, 1992.

71. Gaby, *Preventing and Reversing Osteoporosis,* 83.

72. John R. Lee, *Natural Progesterone,* 40.

73. Balch and Balch, *Prescription for Nutritional Healing* (Garden City Park, NY: Avery Publishing, 1990), 22, and Cedric Garland and Frank Garland, *The Calcium Connection* (New York: G. P. Putnam's Sons, 1988).

74. Garland and Garland, 65–66, 84.

75. Sheehy, *The Silent Passage,* 182.

76. Ibid., 183.

77. The Burton Goldberg Group, *Alternative Medicine* (Puyallup, WA: Future Medicine Publishing, 1993), 776.

78. Ruth Sackman, "Common Sense About Calcium," *Cancer Forum,* Vol. 13, No. 3/4, Fall 1994, 6.

79. John R. Lee, *Natural Progesterone,* 60.

80. Kamen, *Hormone Replacement Therapy.*

81. J. C. Prior, Y. Vigna, and N. Alojada, "Progesterone and the Prevention of Osteoporosis," *Canadian Journal of OB/Gyn & Women's Health Care,* Vol. 3, No. 4, 1991, 181.

82. Ibid.

83. John R. Lee, *Natural Progesterone*, 57-58.

84. Kamen, *Hormone Replacement Therapy.*

85. Trien Susan Falmholtz, *Change of Life* (New York: Fawcett Columbine Books, 1986).

86. Balch and Balch, *Prescription for Nutritional Healing*, 214.

87. Raymond F. Peat, *Nutrition for Women*, 5th ed. (Eugene, OR: Kenogen, 1993).

88. Gaby, *Preventing and Reversing Osteoporosis*, 194.

89. Sydney Lou Bonnick, "Intensive Healing for Brittle Bones," *Prevention*, June 1994, 92–94.

90. Ibid.

91. Gaby, *Preventing and Reversing Osteoporosis*, 234.

92. Bonnick, "Intensive Healing."

93. Gaby, *Preventing and Reversing Osteoporosis*, 234.

94. *The Holistic Dental Digest*, No. 76, June 1992.

95. Danielson et al., "Hip Fracture and Fluoridation in Utah's Elderly Population," *Journal of the American Medical Association*, August 12, 1992.

96. Ibid., and Robert C. Olney, "Stop Fluoride Diseases: Remove Fluorides from Food, Water, Air and Drugs," *Cancer News Journal*, Vol. 9, No. 4, Fall 1955.

97. Nicholas Daflos, "A Tale of Two Federal Agencies," *Cancer Forum*, Vol. 13, No. 3/4, Fall 1994, 9–11.

98. Frederick B. Exner, *The American Fluoridation Experiment.*

99. L. Waldbott, "The Physiologic and Hygienic Aspects of the Absorption of Inorganic Fluorides," *Archives of Environmental Health*, Vol. 2, Issue 2, February 1961, 155–167.

100. Daflos, "A Tale of Two Federal Agencies."

101. Furhman, "Osteoporosis: How to Get It and How to Avoid It."

102. Garland and Garland, *The Calcium Connection*, and Nancy Appleton, *Lick the Sugar Habit* (Garden City Park, NY: Avery Publishing Group, 1988), 14–16.

103. Morter, "Osteoporosis!!" and "The Sodium Connection," *The Chiropractic Professional*, November/December 1987.

104. N. W. Walker, *Fresh Vegetable and Fruit Juices* (Prescott, AZ: Norwalk Press, 1978), 63.

105. Ibid., 63.

106. Ibid., 64.

107. Jean Carper, "As Greens Go, Kale's Best of Bunch," *The Atlanta Journal/The Atlanta Constitution*, August 4, 1994.

108. Gaby, *Preventing and Reversing Osteoporosis*, 116–119.

109. Sally W. Fallon and Mary G. Enig, "Why Butter is Better," *Health Freedom News*, Vol. 14, No. 6, November/December 1995; Earlyne Chaney, *The EYES Have It: A Self-Help Manual for Better Vision* (New York: Instant Improvement, 1993), 38–40; and Nathaniel Mead, "Don't Drink Your Milk," *Health Freedom News*, March 1995, 36–37.

110. Fallon and Enig, "Soy Products for Dairy Products? Not So Fast . . ." *Health Freedom News*, Vol. 14, No. 5, September 1995, Mead, "Don't Drink Your

Milk!", and Chaney, *The EYES Have It*, 38–40.
111. Paavo Airola, *How to Get Well* (Phoenix, AZ: Health Plus, 1985), 190.
112. Ibid., 190.
113. Chaney, *The EYES Have It*.
114. Linda G. Rector-Page, *Healthy Healing: An Alternative Healing Reference* (Sonora, CA: Healthy Healing Publications, 1992).
115. Gaby, *Preventing and Reversing Osteoporosis*, 73.
116. Perlmutter and Hanlon, "Triumph over Menopause."
117. Gaby, *Preventing and Reversing Osteoporosis*, 221–222.
118. Ibid., 220–222.
119. Peat, *Nutrition for Women*, 9, 28.
120. Peat, "Steroids," *Blake College Newsletter*, Vol. 1, No. 4, 3.

CHAPTER 5: THE RISK OF CANCER

1. John R. Lee, *Natural Progesterone: The Multiple Roles of a Remarkable Hormone* (Sebastopol, CA: BLL Publishing, 1993), 74.
2. John R. Lee, "Natural Progesterone" (transcript), *Cancer Forum*, Vol. 13, No. 5/6, Winter 1994-1995.
3. Allen B. Astrow, "Rethinking Cancer," *Lancet*, February 26, 1994.
4. Richard A. Passwater, *Cancer and Its Nutritional Therapies* (New Canaan, CT: Keats Publishing, 1983), 39.
5. McKenna, "Chemicals that Mimic Estrogen May be Causing Reproductive Ills," *The Atlanta Journal/The Atlanta Constitution*, March 8, 1995.
6. John R. Lee, *Natural Progesterone*.
7. Ibid.
8. Lita Lee, "Estrogen, Progesterone and Female Problems," *Earthletter*, Vol. 1, No. 2, June 1991.
9. Ibid.
10. Ibid.
11. Epstein, "The Chemical Jungle: Today's Beef Industry," *International Journal of Health Services*, Vol. 20, No. 2, 1990, 277–280.
12. Julian Whitaker, *Health & Healing: Tomorrow's Medicine Today*, Vol. 3, No. 3, March 1993.
13. John R. Lee, *Natural Progesterone*, 73, 75.
14. Alvin Follingstad, "Estriol, the Forgotten Estrogen?" *Journal of the American Medical Association*, Vol. 239, No. 1, January 2, 1978.
15. John R. Lee, *Natural Progesterone*, 73–75.
16. John R. Lee, "Natural Progesterone," 16.
17. Follingstad, "Estriol."
18. John R. Lee, *Natural Progesterone*, 55.
19. Raymond F. Peat, "Progesterone: Essential to Your Well-Being," *Let's Live*, April 1982.
20. John R. Lee, "Natural Progesterone."
21. Follingstad, "Estriol."

22. Ibid.
23. Ibid.
24. John R. Lee, *Natural Progesterone,* 36.
25. Peat, "Estrogen, Simply Dangerous," *Ray Peat's Newlsetter,* Eugene, OR, July 1995.
26. Passwater, *Cancer,* 24.
27. John R. Lee, "Natural Progesterone."
28. John R. Lee, *Natural Progesterone,* 33, 36, 74.
29. Lita Lee, "Estrogen."
30. Alan R. Gaby, *Preventing and Reversing Osteoporosis* (Rocklin, CA: Prima Publishing, 1994).
31. Peat, *Ray Peat's Newsletter,* Sept. 1996, and R. C. Merrill, "Estriol: A Review," *Physiology Reviews,* Vol. 38, No. 3, 1958, 463–480.
32. Lorraine Day, *Cancer Doesn't Scare Me Anymore!* (videotape) (Rancho Mirage, CA: Rockford Press, 1994).
33. Follingstad, "Estriol."
34. John R. Lee, "Natural Progesterone."
35. Clark et al., "Progesterone Receptors as a Prognostic Facor in Stage II Breast Cancer," *New England Journal of Medicine,* Vol. 309, 1983, 1343–1347.
36. W. L. McGuire, "Steroid Hormone Receptors in Breast Cancer Treatment Strategy," *Recent Progress in Hormone Research,* Vol. 36, 1980, 135–156.
37. Clark and McGuire, "Progesterone Receptors and Human Breast Cancer," *Breast Cancer Research and Treatment,* Vol. 3, 1983, 157–163.
38. John R. Lee, "Natural Progesterone."
39. Ibid.
40. Lita Lee, "Estrogen."
41. Ibid.
42. Betty Kamen, *Hormone Replacement Therapy: Yes or No?* (Novato, CA: Nutrition Encounter, Inc., 1993).
43. Ibid., 122.
44. Ralph W. Moss, *Cancer Therapy: the Independent Consumer's Guide to Non-Toxic Treatment & Prevention* (New York: Equinox Press, 1992).
45. Ibid.
46. Jeff Nesmith, "Breast Cancer Drug Increases Risk . . . ," *The Atlanta Journal/The Atlanta Constitution,* February 22, 1996.
47. Marcus Laux, *Naturally Well,* Vol. 2, No. 12, December 1995.
48. John R. Lee, "Natural Progesterone."
49. "Premarin (Conjugated Estrogens Tablets, USP)," Philadelphia, Ayerst Laboratories, Inc., May 1993.
50. Marcus Laux and Christine Conrad, *Natural Woman, Natural Menopause* (New York: HarperCollins, 1997), 23.
51. Ibid, 21.
52. Anne Dickson and Nikki Henriques, *Women on Menopause* (Rochester, VT, Healing Arts Press), 54.
53. Majid Ali, "Chemical Conflict: The Age of Estrogen Overdrive," *Lifespanner,* December 1994.

54. *Physicians' Desk Reference* (Montvale, NJ: Medical Economics Data Production Company, 1995).

55. "Information for Patients," New York: Ayerst Laboratories, Inc., May 1985.

56. Joe Graedon, *The People's Pharmacy* (New York: Avon Books, 1977), 47.

57. "Dangers of Estrogen" (Form P8263-01), Princeton, NJ: Mead Johnson Laboratories, December 1992.

58. Michael Lerner, *Choices in Healing* (Cambridge, MA: The MIT Press, 1994), 4.

59. Sandra Coney, *The Menopause Industry: How the Medical Establishment Exploits Women* (Alameda, CA: Hunter House, 1994), 93.

60. Ibid., 3.

61. The Burton Goldberg Group, *Alternative Medicine* (Puyallup, WA: Future Medicine Publishing, Inc., 1993), 737.

62. Peat, *Nutrition for Women*, 5th ed. (Eugene, OR: Kenogen, 1993), 17, 21.

63. Burton Goldberg Group, *Alternative Medicine*, 733.

64. Norman C. Shealey, *DHEA: The Health and Youth Hormone* (New Canaan, CT: Keats Publishing, 1996).

65. Walter Pierpaoli et al., *The Melatonin Miracle: Nature's Age-Reversing, Disease-Fighting, Sex-Enhancing Hormone* (New York: Simon & Schuster, 1995), 119.

66. Steven J. Bock and Michael Boyette, *Stay Young the Melatonin Way* (New York: Penguin Books, 1995), 61.

67. Pierpaoli et al., *The Melatonin Miracle*, 27.

68. Ibid., 117.

69. Alan E. Lewis, "Melatonin: Part 2," *Consumer Bulletin, Whole Foods*, March 1996, 102.

70. "Nature's Amazing Healing Oils!" *BioTech News*, 1997.

71. Ibid.

72. Stuart M. Berger, *What Your Doctor Didn't Learn in Medical School* (New York: William Morrow, 1988).

73. Day, *Cancer Doesn't Scare Me Anymore!*

74. Ibid.

CHAPTER 6: HORMONAL SUPPORT FROM OUR FOODS

1. Emanuel Swedenborg, *Arcana Coelestia*, Vol. 8, No. 5949:2, 115.

2. Katharina Dalton, "Premenstrual Syndrome and Postnatal Depression," *Health and Hygiene*, Vol. 11, 1990, 199–201.

3. Raymond Peat, "Diabetes, Scleroderma, Oils and Hormones," *Ray Peat's Newsletter*, Issue 131, July 1995.

4. Egil Ramstad, *Modern Pharmacognosy* (New York: McGraw-Hill, 1959), 119.

5. John R. Lee, "Is Natural Progesterone the Missing Link in Osteoporosis Prevention and Treatment?" *Medical Hypotheses*, 1991.

6. Alan R. Gaby, *Preventing and Reversing Osteoporosis* (Rocklin, CA: Prima Publishing, 1994), xiii.

7. Betty Kamen, *Hormone Replacement Therapy: Yes or No?* (Novato, CA: Nutrition Encounter, Inc., 1993).

8. "Soybeans Ease Menopausal Symptoms," *The Atlanta Journal/The Atlanta Constitution,* April 8, 1993.

9. Ibid.

10. Julian Whitaker, "How Soybeans Delay Aging and Disease," *Health & Healing,* Vol. 5, No. 12, December 1995, 3.

11. Marcus Laux, "Why I Want Women to Choose Natural Hormone Repacement," *Naturally Well,* Vol. 2, No. 12, December 1995, 3.

12. Sally W. Fallon and Mary G. Enig, "Soy Products for Dairy Products? Not So Fast . . . ," *Health Freedom News,* Vol. 14, No. 5, September 1995.

13. Falon and Enig, "Soy Protein."

14. Peat, *Nutrition for Women* 5th ed. (Eugene, OR: 1993), 18, 67, 68, 92.

15. Ibid.

16. Setchell, "Naturally Occurring Non-Steroidal Estrogens of Dietary Origin," *Estrogens in the Environment* (New York: Elsevier Science, 1985), 79–80.

17. Venus Catherine Andrecht, *The Herb Lady's Notebook* (Ramona, CA: Ransom Hill, 1992), 73–76.

18. Laux, *Naturally Well.*

19. Andrecht, *The Herb Lady's Notebook.*

20. Norman C. Shealey, *DHEA: The Health and Youth Hormone,* (New Canaan, CT: Keats Publishing, 1996).

21. Laux, *Naturally Well.*

22. John R. Lee, *Natural Progesterone: The Multiple Roles of a Remarkable Hormone* (Sebastopol, CA: BLL Publishing, 1993), 37.

23. Whitaker, *Health and Healing,* Vol. 2, No. 7, July 1992.

24. Jesse Hanley, "PMS and Menopause Without Pain," *Alternative Medicine Digest,* Issue 10, 1996, 35.

25. Richard A. Passwater, *Cancer and Its Nutritional Therapies* (New Canaan, CT: Keats Publishing, 1983), 25.

26. Fredericks, *Breast Cancer: A Nutritional Approach* (New York: Grosset & Dunlap, 1977).

27. Fredericks, *Guide to Women's Nutrition* (New York: The Putnam Publishing Group, 1989), p 42.

28. Ibid., 30, 31.

29. Fredericks, *Guide to Women's Nutrition.*

30. Ibid.

31. Gail Sheehy, *The Silent Passage: Menopause* (New York: Pocket Books, 1993).

32. Fredericks, *Guide to Women's Nutrition.*

33. Peat, *Nutrition for Women.*

34. Peat, "Diabetics."

35. Peat, "Menopause and Its Causes," *Ray Peat's Newsletter,* Summer 1995.

36. Fredericks, *Guide to Women's Nutrition.*

37. Ibid.

38. Peat, *Nutrition for Women.*

39. Ibid.

40. Peat, "Progesterone: Essential to Your Well-Being," *Let's Live,* April 1982.

41. Passwater, *Cancer.*
42. Ibid.
43. Passwater, *Cancer.*
44. Peat, "Estrogen: Simply Dangerous," *Ray Peat's Newsletter*, July 1995.
45. Fredericks, *Guide to Women's Nutrition.*
46. F. Batmanghelidj, *Your Body's Many Cries for Water* (Falls Church, VA: Global Health Solutions, Inc., 1992).
47. Ibid.
48. Ibid.
49. Ibid.
50. Sam Biser, *Using Water to Cure: The Untold Story* (Charlottesville, VA: The University of Natural Healing, 1994), 28, and Jacques de Langre, *Seasalt's Hidden Powers* (Magalia, CA: Happiness Press, 1993).
51. Earl L. Mindell, "Eggs Have Gotten a Raw Deal," *Let's Live*, September 1994, 8.
52. Peat, *Nutrition for Women.*
53. Sally W. Fallon and Mary G. Enig, "Why Butter is Better," *Health Freedom News*, Vol. 14, No. 6, November/December 1995.
54. Ibid.
55. Fallon and Enig, "Why Butter is Better."
56. Ibid.
57. Peat, *Nutrition for Women.*
58. Peat, "Menopause and Its Causes," *Ray Peat's Newsletter,* Summer 1995.
59. Peat, *Nutrition for Women.*
60. Peat, "Estrogen: Simply Dangerous."
61. Peat, *Nutrition for Women.*
62. Peat, *Progesterone in Orthomolecular Medicine* (Eugene, OR: 1993).
63. Peat, *Nutrition for Women.*
64. James F. Balch and Phyllis A. Balch, *Prescription for Nutritional Healing* (Garden City Park, NY: Avery Publishing Group Inc., 1990), 14, 15, 69.
65. The Burton Goldberg Group, *Alternative Medicine* (Puyallup, WA: Future Medicine Publishing, Inc., 1993), 181–182.
66. Carol H. Munson and Diane K. Gilroy, *The Good Fats* (Emmaus, PA: Rodale Press, 1988), and Udo Erasmus, *Fats & Oils* (Burnaby, BC: Alive Books, 1991).
67. The Burton Goldberg Group Staff, *Alternative Medicine: The Definitive Guide* (Puyallup, WA: Future Medicine Publishing, 1993), and Dorothy R. Schultz, "Why, Tell Me Why?" Hypoglycemia Association, Inc. Bulletin No. 187, August/September/October 1993.
68. Paavo Airola, *Are You Confused?* (Sherwood, OR: Health Plus Publishers, 1971), 76.

CHAPTER 7: PLANNING A PERSONAL APPROACH

1. Christiane Northrup, *Health Wisdom for Women*, Vol. 1, No. 2, February 1995.
2. Gail Sheehy, *The Silent Passage: Menopause* (New York: Pocket Books, 1993).

3. Raquel Martin, *Today's Health Alternative* (Bozeman, MT: America West Publishers, 1992).

4. Sheehy, *The Silent Passage,*

5. John R. Lee, "Significance of Molecular Configuration Specificity—The Case of Progesterone and Osteoporosis," *Townsend Letter for Doctors,* June 1993, 558.

6. John R. Lee, *Natural Progesterone: The Multiple Roles of a Remarkable Hormone* (Sebastopol, CA: BLL Publishing, 1993), 76.

7. John R. Lee, "Significance of Molecular Configuration Specificity."

8. John R. Lee, *Natural Progesterone.*

9. Katharina Dalton, "Ante-natal Progesterone and Intelligence," *British Journal of Psychiatry,* Vol. 114, 1968.

10. Joel T. Hargrove et al., "Menopausal Hormone Replacement Therapy with Continuous Daily Oral Micronized Estradiol and Progesterone," *Obstetrics & Gynecology,* Vol. 73. No. 4, April 1989, 606–612.

11. John R. Lee, "Is Natural Progesterone the Missing Link in Osteoporosis Prevention and Treatment?" *Medical Hypotheses,* 1991.

12. J. C. Prior, Y. Vigna, and N. Alojada, "Progesterone and the Prevention of Osteoporosis," *Canadian Journal of OB/GYN & Women's Health Care,* Vol. 3, No. 4, 1991, 181.

13. John R. Lee, *What Your Doctor May Not Tell You About Menopause* (New York: Warner Books, 1996), 268.

14. Debbie Moskowitz, "New Test Offers Information About Hormone Levels," *Natural Solutions,* Vol. 4, Issue 1, Winter 1996.

15. Marcus Laux and Christine Conrad, *Natural Woman, Natural Menopause* (New York: HarperCollins, 1997), 106, 107.

16. Ibid.

17. Jan Bresnick and Toby Hanlon, "Custom Tailored Hormone Therapy," *Prevention,* July 1995, 65.

18. Alvin H. Follingstad, "Estriol, the Forgotten Estrogen?" *Journal of the American Medical Association,* Vol. 239, No. 1, January 2, 1978.

19. John R. Lee, "Is Natural Progesterone the Missing Link?"

20. Stuart M. Berger, *What Your Doctor Didn't Learn in Medical School* (New York: William Morrow, 1988).

INDEX